ART AND THE EVERYDAY

Art and the Everyday

Popular Entertainment and the Circle of Erik Satie

NANCY PERLOFF

CLARENDON PRESS · OXFORD

1991

Oxford University Press, Walton Street, Oxford OX2 6DP
Oxford New York Toronto
Delhi Bombay Calcutta Madras Karachi
Petaling Jaya Singapore Hong Kong Tokyo
Nairobi Dar es Salaam Cape Town
Melbourne Auckland
and associated companies in
Berlin Ibadan

Oxford is a trade mark of Oxford University Press

Published in the United States
by Oxford University Press, New York

© *Nancy Perloff 1991*

British Library Cataloguing in Publication Data
Perloff, Nancy Lynn.
Art and the everyday : popular entertainment and the circle of Erik Satie/Nancy Perloff.
Includes bibliographical references and index.
1. Satie, Erik, 1866–1925—Criticism and interpretation.
2. Satie, Erik, 1866–1925—Friends and associates.
3. Paris (France)—Popular culture.
4. Popular music—France—Paris—History and criticism.
I. Title.
ML410.S196P47 1991
780'.92—dc20 90–41301
ISBN 0–19–816194–8

Library of Congress Cataloguing in Publication Data
Perloff, Nancy
Art and the everyday : popular entertainment and the circle of Erik Satie.
1. French music. Social aspects
I. Title
780.03
ISBN 0–19–816194–8

Set by Pentacor PLC
Printed in Great Britain by
Biddles Ltd, Guildford and King's Lynn

To Rob

Acknowledgements

AT the University of Michigan's School of Music, where I completed my graduate studies, Glenn Watkins's course on twentieth-century music first stimulated my interest in Erik Satie. Watkins's emphasis on the larger context of twentieth-century art music and the other arts provided an important model as I formulated the thesis which became this book. He directed the entire writing of my original manuscript, always imparting his deep interest and knowledge of the period. Richard Crawford's course on 'Euro-American Folk and Popular Music' prompted my first study of cabaret and café-concert entertainment. While I was writing my paper on French popular entertainment, the likely source of Satie's eccentric humour in the Parisian cabaret occurred to me. Two other members of the University of Michigan faculty, Roland John Wiley and James Dapogny, generously read early drafts of the manuscript. Richard Axsom, a professor of art history at the University of Michigan-Dearborn and a colleague in my second field of study, helped me clarify the complex relation between Cocteau's aesthetic, Apollinaire's *esprit nouveau*, and the Cubist and Futurist movements. It was a special bonus to have the expertise of Jann Pasler, who read large sections of the book when it was near completion and refined specific points of musical analysis as well as the general context of Cocteau's aesthetic.

Many other scholars and specialists helped me grapple with the little-known issues of early twentieth-century concert life and American popular music in Paris. Keith Daniel shared the results of his attempts to locate original versions of early Poulenc works in Paris. Jean-Paul Guiter and Daniel Nevers of RCA-Paris provided me with information on French recording companies and on French records circulating in Paris at the turn of the century. Dominique Raymond of SACEM located French editions of American popular songs. Russell Sanjek, former Vice President for Public Relations of Broadcast Music Inc., offered me information nowhere to be found in print on the various ways in which American musical influence came from West to East. Jacques Fievez, curator of the Collection Théâtre Royal de la Monnaie, Brussels, made available special photographs of the theatre's 1964 reconstruction of *Parade*.

I would like to thank the following institutions: the Bibliothèque and Phonothèque Nationale; the Houghton Library of Harvard

University; the Institute of Jazz Studies, Rutgers University; the Library of Congress; the New York Public Library, including Special Collections and the Rodgers and Hammerstein Archive.

I also wish to thank the Horace H. Rackham School of Graduate Studies of the University of Michigan for a fellowship which greatly facilitated my first year of research and writing.

A wellspring of thanks goes to my husband, Robert Lempert, to whom this book is dedicated. As a specialist in science policy, he offered the invaluable perspective of someone outside the field, taking me to task for inconsistencies and helping me maintain a clear line of argument. His interest in the circle of Erik Satie always refreshed my own.

MUSIC PERMISSIONS

Excerpts from Darius Milhaud 'Marche nuptiale' and Francis Poulenc 'Discours du Général', *Les Mariés de la tour Eiffel* © 1965 by G. Schirmer, Inc. on behalf of Éditions Salabert. International copyright secured. Used by permission.

Excerpts from Francis Poulenc 'Toréador' and Georges Auric *Les Fâcheux* reproduced with the permission of Éditions Salabert, Paris.

Excerpts from Igor Stravinsky *Three Little Songs: Recollections of My Childhood* © by Éditions Russe de Musique; copyright renewed. Copyright assigned to Boosey & Hawkes, Inc. Reprinted by permission.

Excerpts from Georges Auric, *Adieu, New York!*, *Huit Poèmes de Cocteau*; excerpts from Darius Milhaud, *Trois Poèmes de Cocteau*, *Caramel mou*, *Le Bœuf sur le toit*, *La Création du monde*; excerpts from Francis Poulenc 'Le Dauphin' (*Le Bestiaire*), *Cocardes*; excerpts from Erik Satie *Enfantillages Pittoresques*, *Trois Petites Pièces Montées* reproduced with the permission of Éditions Max Eschig, Paris, sole world proprietor.

Excerpts from Henri Christiné 'Valentine' (H. Christiné, Albert Wille-metz) © 1926 Warner Bros., Inc. (Renewed). All Rights Reserved. Used by permission.

Excerpts from Francis Poulenc *Les Biches*; excerpts from Darius Milhaud *Le Train bleu* with the kind permission of Éditions Heugel.

Erik Satie, 'Un Diner à l'Élysée' by permission of the Houghton Library.

Contents

List of Plates   xi

Introduction:
Symphonies without 'Sauce': The Reaction against
     Impressionism   1

    The Formation of the New French Avant-Garde   2
    Jean Cocteau's Anti-Impressionist Stance   7
    Satie's Anti-Impressionist Stance   12
    Anti-Impressionism in the Writings of Milhaud, Poulenc,
     and Auric   14

1. POPULAR INSTITUTIONS IN TURN-OF-THE-CENTURY PARIS   19

    The *Cabaret Artistique*   20
    The Café-concert   24
    The Circus   26
    The Fair   29
    The Music-hall   32
    The Cinema   41

2. THE ARRIVAL OF AMERICAN POPULAR MUSIC AND DANCE ON
     THE PARISIAN SCENE   45

    American Entertainers   46
    Sheet Music and Sound Recordings   59

3. SATIE AND THE CABARET   65

    Accompanist and Composer for Parisian Venues   66
    Cabaret Humour in Satie's Writings   80

4. THE POPULAR WORLD OF COCTEAU, MILHAUD, POULENC, AND
     AURIC   86

    Popular Encounters   86
    The Appeal of Popular Entertainment   97

5. RAGTIME, DIVERSITY, AND NOSTALGIA: THE LANGUAGE OF
     SATIE'S 'PARADE'   112

    Simple Tunefulness   115
     Ostinati and pendulum figures   115

Thematic figures   125
Satie's paraphrase of 'That Mysterious Rag'   132
Musical Diversity   143
Musical Nostalgia   146

6. EMBRACING A POPULAR LANGUAGE   153

Music-hall 'Spoofs', Children's Tunes and Furniture Music,
    1918–1919   153
The 'Spectacle-Concert' of 1920: A Music-hall Re-creation   171
Weapon Against the 'False-Sublime': *Les Mariés de la
    tour Eiffel* (1921)   186
Three 'Monuments to Frivolity' and a Blues Ballet   190
    Les Biches (1924)   191
    Les Fâcheux (1923) and *Le Train bleu* (1924)   198
    La Création du monde (1923)   201
'No Performance': Satie's Last Two Ballets   206

Epilogue   211

Appendix   213

Bibliography   215

Index   223

List of Plates

1. Man Ray, *Erik Satie*, March 1924, photograph Man Ray Trust, Paris.
2. Louis Marcoussis, *Darius Milhaud*, 1936, engraving Musée national d'art moderne, Paris.
3. Man Ray, *Francis Poulenc*, 1922, photograph Man Ray Trust, Paris.
4. Man Ray, *Jean Cocteau*, c.1924, photograph Man Ray Trust, Paris.
5. Pablo Picasso, overture curtain for *Parade*, 1917, tempera on canvas, 10.6 × 17.25 m., Musée national d'art moderne, Paris.
6. Pablo Picasso, costume of the French Manager in *Parade*, painted cardboard, reconstruction (1964) for the Théatre royal de la Monnaie in Bruxelles, Collecton Théatre royal de la Monnaie.
7. Pablo Picasso, costume of the French Manager in *Parade*, 1917, painted cardboard, reconstruction (1964) for the Théatre royal de la Monnaie in Bruxelles, Collection Théatre royal de la Monnaie.
8. Pablo Picasso, costume of the American Manager in *Parade*, 1917, painted cardboard, reconstruction (1964) for the Théatre royal de la Monnaie in Bruxelles, Collection Théatre royal de la Monnaie.
9. *Jardin de Paris*, a poster by Jules Chéret, Musée de la Publicité, Paris.
10. Fernand Léger, stage model of *La Création du monde*, 1923, The Dance Museum, Stockholm.
11. Fernand Léger, king's costume design for *La Création du monde*, 1923, watercolour, 44 × 24 cm., The Dance Museum, Stockholm.
12. Henri Toulouse-Lautrec, *The Trapeze Dancer at the Cirque Médrano*, 1888, charcoal on paper, 37 × 28 cm., Musée Toulouse-Lautrec, Albi.

INTRODUCTION

Symphonies without 'Sauce':
The Reaction against Impressionism

ON 6 June 1917, three weeks after the première of Erik Satie's *Parade*, the poet Blaise Cendrars organized an evening of poetry and music in honour of Satie's ballet. The first performance had triggered one of the most heated scandals of the period, but for the Parisian poets who were Cendrars's guests, *Parade* crystallized ideas they had been exploring since the early 1910s. Many of these poets were friends. Guillaume Apollinaire, who wrote a famous programme note on the ballet, had attended rehearsals of *Parade* with Satie and Jean Cocteau—author of the ballet's scenario. A decade earlier, Apollinaire met Max Jacob and the two spent many evenings at such irreverent cabarets as the Lapin Agile on Montmartre and at the Cirque Médrano. Cendrars adulated his friend Satie, whom he met in 1912 upon returning to Paris from New York.[1] In the minds of all of these artists, *Parade* exemplified a wish to escape Symbolism and fuse 'art' with everyday life. For the exhilarated young musicians Georges Auric, Louis Durey, and Arthur Honegger who joined Cendrars in honouring *Parade*, the ballet symbolized a new direction in French music.[2] They rallied around Satie, who banded with them and announced a new group of musicians which he called 'Les Nouveaux Jeunes'. Although he did not share the youth of his disciples, Satie later explained, he was 'young in character'.[3] Cendrars's event marked the emergence of a new musical avant-garde in Paris.

[1] For the information on Apollinaire and Jacob, see Roger Shattuck, *The Banquet Years: The Origins of the Avant Garde in France, 1885 to World War I*, rev. edn. (New York: Vintage Books, 1968), pp. 294, 263. Material on Cendrars culled from Jay Bochner, *Blaise Cendrars: Discovery and Re-creation* (Toronto: University of Toronto Press, 1978), pp. 42–4, 60.

[2] Satie introduced Auric and Durey to Honegger in 1917. Cocteau and Satie had known one another since 1915 when they began collaborating on *Parade*.

[3] Satie referred to his youthful character in an introductory lecture which he delivered before a concert of 'Les Nouveaux Jeunes' in early 1918. See Erik Satie, *Écrits*, ed. Ornella Volta (Paris: Éditions Champ Libre, 1981), pp. 80–1. Translations here and throughout the book are my own unless otherwise indicated. For an interview in which Satie discussed his membership in 'Les Nouveaux Jeunes', see W. Mayr, 'Entretien avec Erik Satie', *Le Journal littéraire* 24 (Oct. 1924): 11.

Membership in the group grew and then remained constant for several years. As the new aesthetic crystallized, Durey, Honegger, and a more recent member Germaine Tailleferre detached themselves. They admired *Parade*, but were not inspired by Satie's ballet to attack Impressionism and invoke a new French art music based on popular sources. By contrast, three young French composers and members of 'Les Nouveaux Jeunes'—Darius Milhaud, Francis Poulenc, and Georges Auric—joined their mentor Erik Satie and their spokesman Jean Cocteau in fervent recognition of the new cause. Between 1918 and 1924, Satie, Milhaud, Poulenc, and Auric created a repertoire which cast off Impressionist 'vapours' and Romantic sublimity. In place of these elements from the past, the composers infused their music with popular sounds and aesthetic principles (parody, diversity, nostalgia, repetition) derived from Parisian cabaret, circus, fair, and music-hall. Their radical fusion of 'art' and modern life has parallels in the visual arts and in poetry of the period. The startling blend of popular and 'art' idioms which they introduced in French music between 1917 and the early 1920s is the subject of this book.

THE FORMATION OF THE NEW FRENCH AVANT-GARDE

The meeting place of Blaise Cendrars's event was an artist's studio, the Salle Huyghens, owned by the Swiss painter Émile Lejeune. Opened in 1914 as a concert venue which funded a soldier's canteen, it evolved two years later into an interdisciplinary centre for music, art, and poetry.[4] The 'pan-arts festival' that took place between 19 November and 5 December 1916 at rue Huyghens, for example, exhibited paintings by Matisse, Modigliani, and Picasso as well as a programme of poetry by Cendrars and Cocteau. On the pivotal night of 6 June 1917, the musical centrepiece was a two-piano performance of *Parade* by Satie and the Russian pianist Juliette Méerovitch. In addition the programme offered works by each of the young composers: *Trio* by Auric, *Carillons* by Durey, and settings of Apollinaire texts by Honegger. Cendrars, Apollinaire, Jacob, and Cocteau read from their recent poetry.[5] During the next three years, the Salle Huyghens served as one of the principal concert venues for 'Les Nouveaux Jeunes'. With *Parade* as a symbol, membership in the group quickly increased. In the autumn

---

[4] See Bochner, *Blaise Cendrars*, p. 60.
[5] The June 1917 concert programme is outlined by Keith Daniel in *Francis Poulenc: His Artistic Development and Musical Style* (Ann Arbor: UMI Research Press, 1982), p. 13.

of 1917, Cocteau introduced Francis Poulenc, whom he had met through the pianist and teacher Ricardo Vines, and Satie presented Germaine Tailleferre. A seventh composer and a long-time friend of Arthur Honegger's, Darius Milhaud, was a member *in absentia* during the first two years of 'Les Nouveaux Jeunes' (1917–19) when he served as Paul Claudel's secretary in Brazil. The importance of Milhaud's membership is stressed by the singer Jane Bathori, who explains in interviews with Stéphane Audel that Milhaud was present with the composers in thought from the beginning and 'created a true reunion' upon his return.[6]

Satie belonged to 'Les Nouveaux Jeunes' through October 1918. Then, for no clear reason, he withdrew, choosing instead to observe the group's activities and to write about its new aesthetic.[7] Hence membership was fixed at six composers: Auric, Durey, Honegger, Tailleferre, Poulenc, and Milhaud. Between 1917 and 1920 this group gave concerts at two principal performance spots: the Salle Huyghens and the Théâtre du Vieux-Colombier. Events at the Salle Huyghens were directed by the cellist and conductor Félix Delgrange and the singer Pierre Bertin and featured exhibitions of painting as well as music. The subtitle 'Peinture et musique' ('Painting and Music') decorated the concert fliers. Evenings at the Vieux-Colombier offered strictly musical fare and were viewed by its director, Jane Bathori, as a forum for promoting the new group of French composers.[8] Although concerts at both venues interspersed music of Jean Huré, Roger-Ducasse, Roland-Manuel, and other minor French composers with music of 'Les Nouveaux Jeunes', works by the latter tended to appear together on the same programme especially by the second half of 1919. The programme for a Salle Huyghens concert on 5 April 1919, for example, featured premières of Auric's *Trois Poèmes de Jean Cocteau*, Milhaud's String Quartet No. 4, and Poulenc's *Mouvements perpétuels*, as well

---

[6] These interviews were broadcast on Radio-Lausanne in 1953. Transcriptions are housed at the Bibliothèque Nationale as part of the archives of Jane Bathori (originally under the sponsorship of Andrée Tainsy). Even during Milhaud's absence from Paris, concerts of 'Les Nouveaux Jeunes' included his compositions. See, for example, a concert held on 15 March 1918 at the Théâtre du Vieux-Colombier and reviewed by Marcel Orban in *Le Courrier musical* 18 (15 Mar. 1918): 135.

[7] Satie's withdrawal is discussed by Ornella Volta in Satie, *Écrits*, p. 271. The pre-concert lecture delivered in early 1918 (and quoted below) shows that he began to comment on the group's aesthetic before he withdrew.

[8] For a discussion of Jane Bathori's use of the Vieux-Colombier, see Linda Cuneo-Laurent, 'The Performer as Catalyst: The Role of the Singer Jane Bathori in the Careers of Debussy, Ravel, "Les Six", and their Contemporaries in Paris, 1904–1926', Ph.D. Dissertation, New York University, 1982, p. 123. Cuneo-Laurent notes that the composers often performed their own works.

as premières of works by Durey, Tailleferre, Honegger, and Roland-Manuel.[9]

Jean Cocteau was associated with the composers as early as 1917 and shared Cendrars's admiration for Satie. Beginning in November 1916, Cocteau participated in artistic activities at the Salle Huyghens in which he encouraged performances of Satie's music and read poetry he had dedicated to the composer.[10] With the success of *Parade*, a work Cocteau conceived in collaboration with Satie, Picasso, and the choreographer Leonide Massine, he sought the role of spokesman and promoter for 'Les Nouveaux Jeunes'. Cocteau secured this position in 1918 by writing an inflammatory pamphlet, *The Cock and the Harlequin*, in which he identified and interpreted the avant-garde sensibility emerging in French music. Cocteau claimed principally that Satie and his followers were replacing the loftiness of nineteenth-century German music and French Impressionism with a truly French music characterized by brevity, earthiness, and the use of popular sources. In a letter written to Albert Gleizes in the course of completing *The Cock and the Harlequin*, Cocteau expressed his firm stance against Impressionism:[11]

More and more I'm against Impressionist decadence—which doesn't keep me from recognizing the individual values within Impressionism and its unity of style. Yes . . . of course Renoir, but I say down with Renoir the way I say down with Wagner . . . I'm working on *Secteur 131* and on a little book about music . . . Bring me back as many Negro ragtimes and as much great Russian-Jewish-American music as you can.

The first edition of *The Cock and the Harlequin* greeted the Parisian artistic world on 19 March 1918 and rapidly sold out. When Milhaud returned to Paris in 1919, he found that in the aftermath of the war Cocteau had become the manager and Satie the model for an intensely nationalistic movement advocating a 'clearer, sturdier, more precise type of art'.[12] Milhaud not only joined in the ferment but began supplementing the group concerts at the Salle Huyghens and the Vieux-Colombier with Saturday evening dinners at his

[9] This programme is housed in the archives of Jane Bathori at the Bibliothèque Nationale.

[10] Frederick Brown describes Cocteau's involvement in events at the Salle Huyghens in 1916. Among these was a Satie festival to which Cocteau invited Ernest Ansermet and Sergei Diaghilev. See *An Impersonation of Angels: A Biography of Jean Cocteau* (New York: The Viking Press, 1968), p. 158.

[11] Translated and quoted in Francis Steegmuller, *Cocteau: A Biography* (Boston: Little, Brown, & Co., 1970), p. 210. By 'Russian-Jewish-American music', Cocteau probably meant the Tin Pan Alley songs of Irving Berlin and Jerome Kern.

[12] For Milhaud's impressions when he returned to Paris, see Milhaud, *Notes without Music*, trans. Donald Evans, ed. Rollo Myers (London: Dennis Dobson, 1952), p. 80.

apartment, followed by visits to fairground, circus, cinema, and music-hall. On 16 January 1920, after attending one of the private musical gatherings at Milhaud's apartment, the critic Henri Collet wrote an article for *Comoedia* in which he drew parallels between *The Cock and the Harlequin* and writings by Rimsky-Korsakov, between 'Les Nouveaux Jeunes' and the Russian 'Five'. A week later, Collet wrote a second article describing the aesthetic and the music of the six young French composers. This time he christened them 'Les "Six" français'.[13]

By discussing the group of composers and officially designating them 'Les Six', Collet brought public attention to musicians who had been performing privately for three years. Spurred by this new critical acclaim and encouraged by Cocteau, 'Les Six' sought to further their fame and publicity. Between May and November of 1920, Satie joined them in contributing articles and slogans to a broadsheet *Le Coq*, founded and published by Cocteau, Raymond Radiguet, Paul Morand, Lucien Daudet, Max Jacob, and Blaise Cendrars. The tendentious nature of the statements that appeared in *Le Coq* ('The six musicians are no longer interested in harmonic counterpoint', 'We are silly but we are not sublime') and the references to 'Les Six' as the 'Society for Mutual Admiration' suggest that the group deliberately cultivated an iconoclastic tone in order to attract attention and provoke controversy. They successfully achieved this end, moreover, for in the following years a spate of articles appeared in the press, some denouncing 'Les Six' as a hoax and others praising their innovative spirit.[14]

By 1920 the prominence of 'Les Six' as a group force in French music was undeniable. Further collaborative activities took place in 1921 with the production of the *pièce-ballet, Les Mariés de la tour Eiffel*. Yet an important question remains: to what extent did the nationalistic sensibility identified by Cocteau in *The Cock and the Harlequin* represent a shared aesthetic? Over the years, writers on twentieth-century music have associated 'Les Six' with an aesthetic of simplicity and at the same time have observed that Collet's grouping of the composers was in many respects an arbitrary one. Cocteau is partially responsible for this interpretation since, after

[13] See Collet, 'Un Livre de Rimsky et un livre de Cocteau: les cinq russes, les six français, et Erik Satie,' *Comoedia* (16 Jan. 1920): 2; 'Les "Six" français,' *Comoedia* (23 Jan. 1920): 2. This account of the concert which prompted Collet's articles is based on Keith Daniel, *Francis Poulenc*, p. 19.

[14] Émile Vuillermoz denounced Les Six as a publicity stunt in an article, 'The Legend of the Six', *Modern Music* 1, No. 1 (Feb. 1924): 15–19. The English critic Leigh Henry replied with a defence of the group's modernist sensibility. See 'We are Seven', *Modern Music* 1, No. 2 (June 1924): 10–17.

articulating a new movement in French music in 1918, he baff ed
critics by pointing out differences between the composers. Members
of the group added to the confusion, moreover, by insisting that
'Les Six' were bound by ties of friendship rather than a shared
aesthetic.[15]

Careful study of the group's activities and of the individual
composer's views reveals that there was truth in the acknowledgement of differences between them. From 1917 to 1920 Durey,
Honegger, and Tailleferre joined Milhaud, Poulenc, and Auric on
concert programmes, but they did not write personal statements for
Le Coq in which they attacked musical tradition, Wagner, and
Impressionism. Between 1919 and 1921, Durey, Honegger, and
Tailleferre participated in group excursions to Parisian popular
establishments, but they were not nourishing a love for popular
entertainment that began during their youth. Indeed, of the six
composers only Milhaud, Poulenc, and Auric possessed an ardent
admiration for Satie's *Parade* which led them to attack Impressionism and to call for a new French art music. Only these three
composers collaborated individually with Cocteau on works incorporating popular idioms.

On the other hand, Igor Stravinsky did not participate in the
group ventures of 'Les Six' or in the writing for *Le Coq*, but shared
with Satie, Milhaud, Poulenc, and Auric an interest in urban
popular tunes and folk materials and a fondness for musical parody.
As my final chapter demonstrates, parallels for Auric's and
Milhaud's suggestion of 'blue notes', Milhaud's use of bitonality to
imitate the simultaneity of urban sounds, and Satie's re-creation of
children's music appear prominently in Stravinsky's music of the
1910s and early 20s, which also aimed to defy Impressionism. It was
Stravinsky, moreover, who in 1934 hailed Satie's *Parade* for
'opposing the vagueness of a decrepit Impressionism with a precise
and firm language stripped of all pictorial embellishments'. This
admiration was mutual. Satie's long tribute appearing in *Vanity Fair*
in February 1923 praised the 'transparency', the 'clearness of
vibration', the precise 'dynamism' of Stravinsky's music.[16] Yet

---

[15] For instance, in the second issue of *Le Coq* (June 1920), Cocteau defined the periodical
as a broadsheet 'expressing the views of six musicians of different tastes who are united by
friendship'. Similarly, Francis Poulenc explained to Claude Rostand, 'One sees at once that
the Group of Six was not an aesthetic group, but simply a friendly association'. *Entretiens avec
Claude Rostand* (Paris: René Julliard, 1954), p. 46. Milhaud made similar comments in 'The
Evolution of Modern Music in Paris and in Vienna', *North American Review* 217 (April 1923):
550.

[16] Igor Stravinsky, *An Autobiography* (New York: W. W. Norton, 1936), p. 93. Satie's
Vanity Fair article reprinted in *The Writings of Erik Satie*, ed. Nigel Wilkins (London:

Stravinsky's exile in Switzerland during the First World War and his decision to live in France from 1920 to 1929 do not make him a French composer pursuing the strongly nationalistic style favoured by Satie's group. In addition, Stravinsky did not wish to merge 'art' with modern life, nor did his occasional use of parody bear the strong anti-academic, even anti-art stance of Satie and his disciples. For these reasons—and because he did not participate in the group's activities—I have not included Stravinsky in the Parisian aesthetic of the everyday, but will refer to his music wherever it follows principles similar to those of Satie and his colleagues. Such references should broaden the context of art composition in France and its attentiveness to popular music of the day.

During the early 1920s, the close ties between Milhaud, Poulenc, and Auric were noted by the critic Paul Rosenfeld, who also spoke of their adherence to the new aesthetic codified in *The Cock and the Harlequin*. In an article prompted by Durey's resignation from the group in 1921, Rosenfeld asserted: 'With Poulenc, Auric, and Milhaud, we penetrate more closely into the heart of the artichoke. These are the men who carry the group. Without them there would be no Six . . . They have the bite, the courage, the brutality.'[17] The reaction against Impressionism can best be understood by examining the relationship between Cocteau's aesthetic and the views of Satie and his three young disciples.

THE ANTI-IMPRESSIONIST STANCE OF JEAN COCTEAU

As the impresario of a movement against Impressionism, Cocteau derived his aesthetic views not only from the everyday simplicity that he observed in the music of Satie and 'Les Nouveaux Jeunes', but from European literary and artistic currents of the early 1910s. The theoretical basis of Italian Futurism and French Cubism, and the poetics of Apollinaire provided an important source for the attitudes expressed in *The Cock and the Harlequin*.

The tenets of Futurism most influential on Cocteau were those outlined in the manifesto, *The Variety Theatre*, written by the poet

Eulenburg Books, 1980), pp. 102–6. Stravinsky's ballet *Petrushka* (1912) provided an important precedent for the use of urban and folk tunes in art music and for bitonal superimpositions as a means of capturing the fairground medley.

[17] Rosenfeld, 'The Group of Six', in *Musical Chronicle (1917–1923)* (New York: Harcourt, Brace, & Co., 1923), p. 152.

F. T. Marinetti in 1913.[18] Here Marinetti praises principles of speed, instinct, and surprise embodied in the variety theatre and proposes that such aspects of modern urban life be used to rejuvenate contemporary theatre. Marinetti also upholds the variety theatre as a model for artistic change because its disregard for tradition, its celebration of 'swift actuality', and its recourse to a 'simplicity of means' destroy the 'Solemn, the Sacred, the Serious, and the Sublime in Art with a capital A'.

The 'Solemn, the Sacred . . . the Sublime' were precisely the qualities cherished by the Symbolist poet Stéphane Mallarmé as testaments to the purity and refinement of art and, specifically, of poetry. An essay, 'As for the Book', which Mallarmé wrote in 1895, epitomizes the late nineteenth-century artistic principles which Marinetti opposed in 1913. In a section of this essay entitled 'The Book: A Spiritual Instrument', Mallarmé urges writers to preserve the book as a 'hymn' and a 'divine and intricate organism'. The purity of the book can be maintained, he argues, by keeping its discourse lofty and remote from the quick, unrefined writing of common newspapers.[19]

Mallarmé's exalted view of art was questioned by the Cubist painters Picasso and Braque as well as by the Futurist poet Marinetti. Cubism was the first of the two movements, moreover, to defy the traditional division separating the external reality of nature from the internal reality of art.[20] In 1909–10 in his *Still Life with Violin and Pitcher*, Georges Braque painted a 'trompe-l'oeil' nail on the top of the canvas. By casting a shadow, just as a real nail would, on the flat surface of the canvas, Braque's nail shattered the principle—fundamental in Western painting since the Renaissance —that the picture surface is a transparent surface offering an illusion of reality. Two years later Picasso went further. In his *Still Life with Chair Caning* (1911–12), he pasted a strip of oilcloth on the lower surface of the canvas rather than painting it. This gesture, which introduced the first collage, rejected the Western convention that an illusion of reality must be achieved with pencil and paint alone.

The challenges posed by Futurism and Cubism to Mallarmé's ideal of a spiritual book were echoed by Apollinaire. In 1912,

[18] For a translation, see F. T. Marinetti, 'The Variety Theatre', in *The Documents of Twentieth-Century Art: Futurist Manifestos*, ed. Umbro Apollonio (New York: The Viking Press, 1970), pp. 126–31.
[19] For a translation, see Mary Ann Caws, ed., *Stéphane Mallarmé: Selected Poetry and Prose* (New York: New Directions Publishing Corps., 1982), pp. 80–4.
[20] The following discussion of Cubist innovations is based on Robert Rosenblum, *Cubism and Twentieth-Century Art* (New York: Harry N. Abrams, 1976), pp. 45, 68.

inspired by the appearance of the first Cubist collages, Apollinaire began modelling his poetry on the subject-matter and the typographic layout of newspapers, advertisements, and posters. In the October 1912 issue of *Soirées de Paris*, a recently founded review for which Apollinaire was associate editor, he explained: 'For some time I have been trying new themes far different from those around which you have seen me entwine my verses up to now. I believe that I have found a source of inspiration in prospectuses . . . catalogues, posters, advertisements of all sorts. Believe me, they contain the poetry of our epoch. I shall make it spring forth.' In June 1914, Apollinaire introduced the first poems he called 'calligrammes', consisting of fragments of speech arranged spatially across the page. The aesthetic principle behind these visual-verbal patterns was 'simultanism', a term coined by Apollinaire to mean the 'opposite of narration', the attempt to capture the moment of experience in which all parts interpenetrate through contrast, rather than through logical discourse.[21] Darius Milhaud later took up the simultaneity principle when he superimposed three or more tonalities in a re-creation of his surrounding urban environment. For Apollinaire, the adoption of advertisements and the subsequent invention of the calligram marked an effort to escape Symbolism and close a gap that Mallarmé urged poets to keep open: that between poetic and everyday discourse.

The conscious attempt in Futurism, Cubism, and the poetics of Apollinaire to break from hallowed traditions of the past and to fuse art with commonplace materials formed the crux of their influence on Cocteau's *The Cock and the Harlequin*. For his specific critique of Impressionism, Cocteau relied heavily on Jacques Rivière's historic review of *The Rite of Spring* published in *La Nouvelle Revue française* in November 1913.[22] Hailing Stravinsky's ballet as the first masterpiece to oppose Impressionism, Rivière provided the first definition of an anti-Impressionist style. He argued that the great novelty of *The Rite of Spring* was its renunciation of the 'sauce' and its adoption instead of 'rawness', 'purity', 'clarity', as well as 'contours' never blurred by 'veils' or other 'vaporous quiverings'. Rivière praised Stravinsky's decision to exclude the evocative, expansive sound of strings in favour of dry, clear woodwinds. With

---

[21] For references to Apollinaire's interest in posters and to his invention of calligrams, see Shattuck, *The Banquet Years*, pp. 279, 286–7, 310–11.

[22] The following discussion taken from Rivière, 'Le Sacre du printemps', *La Nouvelle Revue française* 10, No. 59 (Nov. 1913): 706–8, 714–15. I am indebted to Jann Pasler for pointing me to Rivière's review as a major influence on Cocteau's definition of anti-Impressionism.

woodwinds, he argued, Stravinsky's statements sound direct, even compressed so that a new brevity emerges. At the beginning of his review, Rivière stressed that the creator of the spurned Impressionism was Claude Debussy: 'Without violence, without ingratitude, but very clearly, Stravinsky disengages himself from Debussyism. He has understood that this delicious halo amidst which his master's music always seems to be drowning, risks becoming nothing more than sauce in the hands of a disciple. He deliberately removes all indecision, all trembling from his symphony.' Rivière focused throughout on Stravinsky, but attributed the precision of *The Rite* to a tendency typical among Russian composers 'from the beginning' to uphold distinctness and reject 'effusion' and 'flowing atmospheres' in their music. He referred here to Mussorgsky and Stravinsky, rather than to Rimsky-Korsakov with his 'exotic picturesque' style.

In *The Cock and the Harlequin* Cocteau paraphrased Rivière's argument on the virtues of anti-Impressionism and the dangers of Impressionism. He also borrowed the Futurist technique of presenting these ideas in a manifesto which resembled an advertisement in its use of bold, startling slogans. The intention lay as much in attracting publicity as it did in winning people over to a new cause— one that directly challenged Rivière's eulogy of Stravinsky's anti-Impressionism.

The cause was a new French music based on the style of Erik Satie. According to Cocteau's retelling of music history, Satie represented the first truly French composer of the twentieth century; the Impressionist movement, epitomized by Debussy, was not really French at all. In order to make this argument convincing, Cocteau defined Impressionism as a 'misty haze' initiated by the German Romanticism of Wagner and furthered by Mussorgsky, Debussy, and Stravinsky. The Impressionist musicians 'thought the orchestra in *Parade* poor', he declared, 'because it had no sauce'.[23] It was Cocteau's view that Wagner had set a dangerous precedent for musical 'sauce' by writing subjective, densely textured pieces that devoured an audience just as an 'octopus' devours its prey. Cocteau considered the atmosphere at Bayreuth a 'religious complicity among the initiated' and Wagner's music dramas works which 'hypnotized the audiences at Bayreuth . . . and plunged them into semi-darkness'. Such a view reflected the same distaste for artistic

---

[23] Quotes in this and the following paragraph are taken from Jean Cocteau, *The Cock and the Harlequin*, in *A Call to Order*, trans. Rollo Myers (New York: Haskell House, 1974), pp. 16, 24, 35.

spirituality that inspired the writings of Marinetti and Apollinaire.

In Cocteau's attack on Impressionism, Debussy and Stravinsky were coupled with Wagner because both had inherited the Romantic tradition of writing music that one 'listens to through one's hands'. Debussy's music was tainted by the Wagnerian 'fog' and by the Russian 'fire'—that is, the rich orchestral sonorities—of Mussorgsky. Stravinsky's produced the same atmosphere of 'religious complicity' as Wagner's. Hence all three belonged to the category of 'hazy' music which constituted Impressionism. The accuracy of this association of Debussy and Stravinsky with Wagner is of course highly suspect, given Debussy's attempts to escape the 'ghost of old Klingsor' and Stravinsky's consistent stance against Wagner. As recently as 1914, moreover, Cocteau had endorsed Stravinsky and approached him for collaboration on a ballet, *David*. Stravinsky did not pursue the project, and Cocteau was greatly disappointed. In *The Cock and the Harlequin*—partly to avenge himself but also to wage a new campaign—Cocteau grouped Stravinsky with Wagner and rewrote music history so that Satie appeared unique among contemporary composers. The language Cocteau used in much of his manifesto was identical to the language Rivière used five years earlier in his review of *The Rite*—but to make a very different point. Rivière cast Stravinsky as the hero of anti-Impressionism, while Cocteau cast Satie in this role and made Stravinsky the Impressionist villain.

If Cocteau tended to lift words and even entire arguments from Rivière's discussion of Impressionism, his explanation of Satie's anti-Impressionist qualities was, for the most part, original. An important example was Satie's defiance of the artist as genius. Whereas Impressionism, Cocteau argued, carried with it a nineteenth-century belief in the genius who defines an epoch, the modest tuneful music of Satie resisted deification and offered a 'clear road open upon which everyone was free to leave his own imprint'.[24] Whereas the Impressionists employed a lofty, long-winded discourse, Satie wrote concise music inspired by practices of cabaret and music-hall. According to Cocteau, Satie's emphasis on melodic simplicity and clarity and his opposition to a treatment of art as sacred and inviolate marked the burgeoning of a French music freed from German Romantic and Impressionist influences.[25]

[24] *The Cock and the Harlequin*, p. 18. For a statement in which Cocteau warned of the dangers of 'masterpieces' which, like a domineering composer, encourage only fawning schools of imitators, see 'La Jeunesse et la scandale,' in *Œuvres complètes* (Geneva: Marguerat, 1946) ix: 336, 340.

[25] Clarity and concision were of course points also emphasized by Rivière.

SATIE'S ANTI-IMPRESSIONIST STANCE

Years before Cocteau wrote *The Cock and the Harlequin*, Satie attacked pedantry, academic seriousness, and artistic sublimity in his many essays and in the humorous performance directions adorning his piano pieces. An autobiographical sketch written in 1913 deflated the Romantic view of the artist by making sarcastic, intentionally pompous references to Satie as 'the precious composer' and the 'master' of 'lofty genres'. In a parody of academic guidelines for musical composition entitled 'Conservatory Catechism' (1914), moreover, Satie mocked the adulation which young composers in France lavished upon Debussy's advanced melodic and harmonic techniques. The first commandment read: 'Dieubussy alone shalt thou adore. And copy most perfectly', and the following eight instructed one to systematically employ all the new practices introduced by Debussy—parallel fifths, unresolved dissonances, ninth chords, open form. Implicit in this satiric codification of Debussy's style was Satie's belief that the innovations of Debussy went hand in hand with a nineteenth-century image of the deified composer. Also in 1914, Satie taunted pedants (whom he called the 'dried up' and the 'stultified'), by writing a 'Preface' to *Sports et divertissements* in which he noted that he had tried to make the opening chorale in the collection as boring as possible.

Cocteau may not have been familiar with these early writings and annotations but he was certainly aware of the irreverent attitude, since Satie offered the same sarcastic mockery of official trappings during musical gatherings of 'Les Nouveaux Jeunes'. In early 1918, for example, Satie introduced a concert at the Théâtre du Vieux-Colombier with the words, '[Les Nouveaux Jeunes] have neither President nor Treasurer, Archivist nor Economist. We don't have any treasure either. That is handy for us.'[26] Clearly, then, Cocteau's *The Cock and the Harlequin* was as influenced by the aesthetic stance of Satie as by the everyday nature of *Parade*. The influence was reciprocal, moreover, for Satie's essays attacking academicism appeared much more frequently after Cocteau had published his manifesto in 1918. In addition, Satie's newly acquired fame gave him the impetus to draw a firm link between Impressionism and nineteenth-century notions of tradition and the sublime. This link resonates in the writings of Cocteau and Satie's three disciples.

In his 'Conférence sur les Six', for example, written in 1921, Satie celebrated the modern sensibility of the six young composers and

---

[26] See Satie, *Écrits*, ed. Volta, p. 80. For the quotes taken from Satie's autobiographical sketch and the Conservatory Catechism, see *The Writings of Erik Satie*, pp. 79, 81.

dismissed Impressionism as a movement of the past. 'I was myself, thirty years ago, terribly "impressionistic", he observed.[27] In an article that appeared in *Le Coq* in June 1920, moreover, he offered a vehement attack on the principle of schools and disciples which he saw represented in Impressionism. Echoing Cocteau's claims in *The Cock and the Harlequin*, Satie warned that schools of musical composition were dangerous because they encouraged a kind of bondage in which disciples slavishly imitated the techniques of one composer. The result, in Satie's view, was a devastating reuse of the same musical mannerisms, so that what had once appeared novel was now tediously and predictably formulaic. In his case, however, Satie argued that no school would develop because he would not allow it: 'I never attack Debussy. It's only Debussyites that annoy me. THERE IS NO SCHOOL OF SATIE. Satieism would never exist. I would oppose it. There must be no slavery in art. I have always tried to throw followers off the scent by both the form and content of each new work.'[28] Satie concluded his article by praising Cocteau for having helped the young musicians move away from the 'boring, provincial and academic mannerisms' of the latest Impressionist music.

Satie may have echoed Cocteau's attacks on the imitative, formulaic Debussyites, but he diverged from the poet in his genuine respect for Debussy. An article written for *Vanity Fair* in 1923, five years after Debussy's death, contained tongue-in-cheek references to the positions of Debussy and Wagner as artists whose names will never vanish from the 'Universal Memory . . . for Glory has inscribed [them] forever in the splendid and most rich Honours List of Music'.[29] Yet beneath the mockery of artistic legacies lay a sincere tribute to a composer with whom Satie had felt a perfect understanding. In his article, Satie also assumed the responsibility for having guided Debussy away from Wagner and from 'sauerkraut', towards a simpler style. For Satie, the Debussyites were the enemy. They would have caused a separation between Satie and Debussy, had Debussy still been alive.

Satie's writings reveal that anti-Impressionist sentiments harboured before the appearance of *The Cock and the Harlequin* were expressed more brazenly in the years that followed. Hence Satie inspired Cocteau, who in turn directed Satie. The attitudes of Satie

[27] Satie, *Écrits*, ed. Volta, p. 90.
[28] This translation of Satie's article appears in *The Writings of Erik Satie*, p. 84. Beginning with *Pulcinella*, the principle of throwing followers off the track with each new work was taken up by Stravinsky, and can be traced to Diaghilev's injunction to Cocteau, 'Astonish me!'
[29] See *The Writings of Erik Satie*, p. 106.

and the theories of Marinetti, Picasso, Apollinaire, and Rivière make it apparent, then, that *The Cock and the Harlequin* reflected the influence of several contemporary calls for a break from the nineteenth-century German tradition and from Impressionism. The choice of Erik Satie as the model of French modernism and the celebration of circus and music-hall as inspirations for French composers were Cocteau's own creation.

ANTI-IMPRESSIONISM IN THE WRITINGS OF MILHAUD, POULENC, AND AURIC

With Milhaud, Poulenc, and Auric, it is also difficult to trace a single source of influence on their revolt against Impressionism. More appropriately, one might argue that the similarity between their objections to Impressionism and those of Cocteau and Satie reveals the complex interweaving of attitudes shared by a group of musicians and artists who met and performed together.

Even before the première of *Parade*, Georges Auric valued the music of Satie. In an essay written for *Le Courrier musical* in March 1917, he praised Satie's spirit of farce and his diffidence and advised readers to pay close attention to the musical activities of this 'good master'.[30] With the first production of *Parade* and the publication of Cocteau's manifesto a year later, Milhaud and Poulenc joined Auric in celebrating the new ballet which, in their eyes, made the crucial break with the Impressionist and Romantic past. An article written by Milhaud in 1923, for example, welcomed Satie's re-establishment of musical simplicity and described how 'the souvenir of music-halls' in *Parade* 'creates a new and attractive art'.[31] Auric's review of a second performance of *Parade* in 1921 praised the jazz and circus world of Satie's ballet because it so effectively counteracted the 'clouds and sirens of Debussyism'.[32] And in interviews with Stéphane Audel in 1963, Poulenc recalled how in 1917 he was 'conquered by *Parade*' and willing to disown Debussy, so eager was he for the 'new spirit Satie and Picasso were bringing us'.[33] Poulenc's alignment of Satie and Picasso here suggests his awareness that the two artists shared similar aims: both employed materials of everyday life (newspaper, cloth, cardboard in the case

---

[30] *Le Courrier musical* 18 (1 Mar. 1917): 129–30.

[31] Darius Milhaud, 'The Evolution of Modern Music in Paris and in Vienna', p. 549.

[32] Georges Auric, 'Les Ballets-Russes: à propos de *Parade*,' *La Nouvelle Revue française* 16 (1921): 224.

[33] Francis Poulenc, *My Friends and Myself*, conversations assembled by Stéphane Audel, trans. James Harding (London: Dennis Dobson, 1978), p. 39.

of Picasso, popular tunes and repetition in the case of Satie) in order to challenge traditional aesthetic criteria and to make art self-referential.[34]

Statements championing *Parade* were accompanied by essays, interviews, and aphorisms in which the three young composers lashed out at Impressionism. For Milhaud, Poulenc, and Auric, Impressionism was an ethereal, overrefined musical writing practised by the imitators of Debussy (the Debussyites). Milhaud and Poulenc joined Satie in preserving Debussy from attack. Milhaud was the only one of the three, however, to discuss the origins of Debussyism, and here he aligned his argument more closely with Satie and Rivière than with Cocteau. For instance, when he identified the 'Russian snare' as the source of Impressionism, Milhaud aimed his criticism at the exotic orchestral techniques of Rimsky-Korsakov which entrapped French music in 'an alley of rare and delicate sonority'. Like Rivière, he praised the direct, pure sounds of Mussorgsky and Stravinsky because they caused a beneficial reaction against Impressionism. Naturally Milhaud lavished his greatest praise on Erik Satie whom he credited with anticipating the dangers of the Impressionist school and changing his course so as to recover the 'real French tradition'. After studying counterpoint and fugue, Milhaud recounts, Satie began writing pieces 'intended to be satirical and directed against Impressionist "poetry above all"'. Satie then successfully counteracted 'the blow of *The Rite*' and brought French music back to simplicity. Milhaud's criticism here of Stravinsky—whom he in fact admired—reveals the core of French nationalism which lay beneath his objections to Impressionism.[35]

All three young composers assailed the notion of masterpieces, which they believed Impressionism had inherited from the nineteenth century. In the May 1920 issue of *Le Coq*, Poulenc warned: 'We will never give you "works" ['Nous ne vous donnerons jamais d'œuvres']. Milhaud elaborated on this statement in his *Études*, explaining that by 'works' Poulenc meant the long symphonic poems of the mid-to-late nineteenth century, which were laboriously developed and densely orchestrated.[36] Georges Auric delivered

---

[34] For further discussion of Picasso's Cubist aims in *Parade*, see Richard H. Axsom, *'Parade': Cubism as Theater* (New York: Garland Publishing, 1979), pp. 175–88.

[35] Quotes in this paragraph from Milhaud, 'The Evolution of Modern Music', pp. 548–9. Milhaud referred to the 'real French tradition' in conversation with John Alan Haughton during his trip to the United States. See Haughton, 'Darius Milhaud: A Missionary of the Six,' *Musical America* 27 (13 Jan. 1923): 3.

[36] Milhaud, *Études* (Paris: Éditions Claude Aveline, 1927), p. 27.

similar slaps in the face of musical tradition when he predicted that *Parade*, with its directness and its basis in popular music, would prepare a 'new world' in which the old 'works' of the nineteenth century would no longer be necessary or important.[37]

Alongside this criticism of nineteenth-century notions of master-works, Poulenc disparaged such Debussyites as Florent Schmitt and Maurice Ravel for playing the role of disciples and following a 'fad'. 'I'd had my fill of whole-note scales, harp glissandos, muted horns, harmonics, and the sound of strings', Poulenc wrote, outlining his conception of Impressionism as he attempted to distinguish himself from these imitators of Debussy.[38] Poulenc's distaste for musical systems and schools of composition prompted him to invoke instinct as his surest guide and to embrace artistic novelty. In conversations with Stéphane Audel, he referred to Picasso's statement, 'Long live disciples! It's because of them that we go looking for something else', as a perfect expression of his own iconoclastic sentiments during the years following *Parade*. Satie was an ideal leader, moreover, because, according to Poulenc, he encouraged the young composers to write and perform together without forcing his leadership on the group.[39] Such assertions of artistic independence may explain why Poulenc denied that a common aesthetic bound the 'Six'.

Milhaud aimed his criticism of disciples and schools principally at Wagner's domination of French music. Throughout his writings, he suggested that 'the next phase in the development of our music' could take place only if French composers rid themselves of previous influences—Russian and Debussyian as well as Wagnerian. Indeed the 'outworn idioms' and 'sentimental harmonies' of the preceding phase, Impressionism, reflected the predicament of French composers unable to escape these influences.[40] Hence Milhaud championed the new and employed a deliberately brash language in order to make his point. His decision to replace his music review of the Concerts Pasdeloups with the statement 'Down

[37] Auric, 'Les Ballets-Russes: à propos de *Parade*', p. 224. The attacks on 'masterpieces' appeared most prominently in the composer's youthful writings (see also Milhaud's criticism of Wagner). Biographical statements written in the 1950s and later testify to the receding importance of such claims once the composers had established their own reputations.

[38] Poulenc, *My Friends and Myself*, p. 64.

[39] Poulenc's statement on disciples appears in *My Friends and Myself*, p. 127. For his praise of Satie's style of leadership, see Poulenc, *Entretiens avec Claude Rostand*, p. 46.

[40] For these quotes, see Milhaud, *Notes without Music*, p. 26. Milhaud's conviction that the 'Six' must free themselves from the influence of Wagner and the preceding French generation is captured in the line, 'Duparc, d'Indy, and Chausson were academic to Ravel and Debussy, and they in turn are academic now'. Quoted in Haughton, 'Darius Milhaud: A Missionary of the "Six"', p. 3.

with Wagner' and his eagerness, upon returning from Brazil, to join his Parisian colleagues 'who stood for the unshackling of art' are just two examples of this manner.

As the youngest member of the group, Georges Auric launched the most daring and all-encompassing attacks on Impressionism, which in his case subsumed Stravinsky and Debussy as well as their followers. His arguments thus bore the closest resemblance to Cocteau's. They centred on Impressionist 'sublimity'. Auric used the word 'sublime' to mean artistic vanity and the overpowering presence of the composer in every chord, note, and measure of a musical work. In *Le Coq* (June 1920), he made the following comments: 'Debussy, I know, reached the sublime in his throbbing chords. I find the sublime on every page and in every measure of Stravinsky's ballets. How could we forget *The Rite of Spring*, that extraordinary tumult—an entire orchestra dominated by the genius of a man who for us was his whole race that day.' The sarcastic reference to 'the sublime on every page' was a deliberate negation and reversal of a praiseworthy comment made by Jacques Rivière in his review of *The Rite*: 'One will find grace on each page of *The Rite of Spring*, in these profile faces resting on the shoulders in front . . . in the dark, sparse, absorbed promenade of the Adolescents in the Second Tableau.'[41] Since, in Auric's view, sublime music resembled religion in its power to hypnotize audiences and plunge them into a dark mist, criticism of the sublime comprised an attack on the same principle of masterworks and artistic schools decried by Cocteau, Satie, and Auric's two young colleagues. All four advocated an escape from forces that were complicating French music, and a turn to a more objective, modest art.

In the essays and aphorisms of Satie, Cocteau, and the three younger composers, the deliberate donning of a witty, satiric pose and the use of provocative language makes it difficult to know how seriously the five artists wished to be taken. Critics like Émile Vuillermoz and René Dumesnil, writing in the 1920s, declared that the 'Six' had cultivated scandal and invented a legend about their radical techniques in order to win fame and dupe the public.[42] Vuillermoz described them as a 'commercial firm', a publicity stunt. Although it is certainly true that Cocteau, Satie, and their young colleagues took advantage of the attention which they received from

---

[41] Rivière, 'Le Sacre du printemps', p. 726.

[42] See Émile Vuillermoz, 'The Legend of the Six', and René Dumesnil, *La Musique contemporaine en France*, 2 vols. (Paris: Librairie Armand Colen, 1930), ii: 11. Dumesnil wrote: 'As Vuillermoz said, what seemed so irritating about the Six was that they consecrated themselves apostles of a new faith. Thus the ignorant public . . . remained persuaded that the Six had invented a new style.'

Henri Collet and wrote bold, inflammatory statements, an icon-oclastic tone was intimately bound up with the determination to break from the German Romantic past. Not only the language but the forums in which the group expressed its views—a broadsheet, *Le Coq*, and a manifesto, *The Cock and the Harlequin*—had a distinctly modernist flavour and hence embodied the very rupture with tradition which they were demanding. By modelling their music on Parisian popular entertainment—aspects of which had inspired Satie's *Parade*—Milhaud, Poulenc, and Auric sought to establish a modern French tradition based on the rich assortment of popular milieux, genres, entertainers, and brands of humour flourishing in early twentieth-century Paris.

Cocteau, Satie, and the three younger composers based their conception of an aesthetic of the everyday on close personal contact with *fin-de-sièle* Parisian popular entertainers and milieux. In my first two chapters, I will consider the array of Parisian institutions, the types of French and American entertainers who performed there, and the assortment of popular genres comprising the repertoire. The picture that unfolds will provide a backdrop for a study in the next two chapters of the composers' encounters with popular performers and venues and their embrace of certain aesthetic principles associated with music-hall, fair, and circus (parody, diversity, simplicity, nostalgia). My final two chapters will offer a musical analysis: in Chapter 5 of the popular language of Satie's celebrated ballet *Parade* and in Chapter 6 of the art music composed by Milhaud, Poulenc, and Auric between 1918 and 1924, which drew its inspiration from vernacular sources and from Satie's treatment of diversity, simplicity, and nostalgia in *Parade*.

I

Popular Institutions
in Turn-of-the-Century Paris

IN a letter to Francis Poulenc, dated 2 September 1918, Jean
Cocteau expressed his love for the magical world of Parisian popular
amusement: 'Are you familiar with the fair of Bordeaux and the
camouflaged ships? Even better, the Spectacle Casino de Paris.
Merry-go-rounds dizziness world upside-down velvet mirrors and
enamel-painted Louis XIV horses which are rearing in a paradise of
dentists and theatre loges.'[1] The jumble of entertainment spots in
this passage (fair, music-hall, merry-go-round) is significant because
it suggests that Cocteau was enchanted by an environment which
crossed the bounds of individual milieux and dazzled the audience
with its rapid succession of diverse images and its buoyant mood.[2] In
their writings, Milhaud, Poulenc, and Auric displayed the same
willingness to embrace the spectrum of Parisian popular entertainment. Their descriptions of Saturday evening visits to popular
haunts are charged with a delight in being immersed in the din of
merry-go-rounds, shooting galleries, and revue tunes and in the
comedy of clown acts. Although Satie was less explicit, his work
as accompanist for cabaret, café-concert, and music-hall brought
him into close contact with popular institutions, and ballets such
as *Parade, Mercure*, and *Relâche* reflect the absorption of many
different French and American popular sources.[3]

For Cocteau and the four composers, then, the different Parisian
popular establishments tended to merge into one vast entertainment
world. Such a view had a firm basis in reality; exchanges between

[1] Francis Poulenc, *Correspondance 1915–1963*, compiled by Hélène de Wendel (Paris:
Éditions du Seuil, 1967), p. 19. The Casino de Paris was a popular music-hall. Camouflaged
ships may be a reference to a game at the fair of Bordeaux.
[2] Several years earlier Cocteau incorporated the same dizzy succession of images into his
character descriptions for *Parade*. The description for the Little American Girl is included in
Frederick Brown, *An Impersonation of Angels* (New York: The Viking Press, 1968), pp. 128–
9.
[3] The sense of delight in being immersed in a wide variety of entertainment spots emerges
with particular vividness in Milhaud's account of his Saturday evening dinners and the group
visits to popular haunts. See Darius Milhaud, *Notes without Music*, trans. Donald Evans
(London: Dennis Dobson, 1952), pp. 83–4.

institutions and diverse programmes at individual amusement spots were facts of life in late nineteenth- and early twentieth-century Paris. Indeed the diversity of turn-of-the-century Parisian popular venues reflected the startling simultaneity of modern urban technological life, with its intricate network of patterns, sights, and sounds. My account of the popular world of Satie, Milhaud, Poulenc, Auric, and Cocteau will move from institution to institution, addressing in each case the complex interchange and the highly eclectic repertoire. In an attempt to explore popular milieux through the eyes of the five artists, particular attention will be given both to the diversity and simultaneity and to the parody and humour which inspired the anti-Impressionist aesthetic.

THE *CABARET ARTISTIQUE*

Since the mid-eighteenth century, Parisian cafés and salons had offered a forum for artists and writers to meet and exchange ideas. Salons, moreover, hosted dramatic recitations by famous professional actors. The years of peace which lay between the Franco-Prussian War and World War I witnessed the decline of the salon and the shift of actors to the café as a performance arena.[4] Coupled with this development was the emergence of a new establishment— the *cabaret artistique*—which featured song as well as theatre and provided a counterpart to the literary cafés of Paris.

The Chat Noir, founded in 1881 by Rodolphe Salis, was the first *cabaret artistique* to achieve fame. Salis emphasized the 'artistic' tone of his establishment by decorating it in the style of Louis XIII, with murals, muskets, swords, caskets, cups reputed to have belonged to Charlemagne, and other medieval icons adorning the walls.[5] He also devised an evening programme which suited his literary pretensions. Typically, during the cabaret's early years, this programme began with a rousing chorus performed by a group of poets and singers and conducted by one member of the company. From there, the cry of 'A la tribune!' signalled the first in a series of performances by individual *poètes-chansonniers* (poet/song-writers) —some beginners, others more experienced—who stepped up to the counter and displayed their talents in song- and poetry-writing,

---

[4] Roger Shattuck discusses the decline of the salon in *The Banquet Years: The Origins of the Avant-garde in France, 1885 to World War I*, rev. edn. (New York: Vintage Books, 1968), pp. 10–11.

[5] The artistic décor of the Chat Noir was designed in imitation of an earlier cabaret, la Grande Pinte. See Michel Herbert, *La Chanson à Montmartre* (Paris: Éditions de la Table Ronde, 1967), pp. 61, 64. For further information on the evening programme and the atmosphere at the Chat Noir, described in this paragraph, see also *La Chanson à Montmartre*, pp. 69, 72–3.

singing, and dramatic recitation. Each performer was introduced by the poet and master-of-ceremonies, Émile Goudeau, while Salis poured the beer and waiters dressed as academicians served the drinks. The atmosphere at the Chat Noir was noisy and informal; jokers called *fumistes* interjected puns, witty lines, and humorous tales in between songs.

In his memoirs, the poet and singer Maurice Donnay evokes the mood of these early cabaret performances and identifies attitudes and sentiments considered 'artistique' and therefore cultivated by *poètes-chansonniers* of the time:[6]

One would not know how to imagine, in 1881 in a Louis XIII cabaret in Montmartre, what this word 'artist' could encompass in terms of youth, gaiety, audacity, lyricism, fantasy, an I-could-care-less attitude, a certainty in the uncertainty of tomorrow, a subversive theory of *fumisterie*, a smoke of glory, a smoke of tobacco, a thirst, a taste for beards and long hair.

The bohemian attitude described by Donnay went hand in hand with a penchant for satire and hyperbole which characterized both the cabaret songs and the anecdotes appearing in the weekly journal of the Chat Noir. For instance, in the 9 May 1885 issue of this journal, a dramatic announcement of the Chat Noir's upcoming move offered a witty parody of sensationalist newspaper reporting. 'From 15 May to 20 May in the year of grace 1885', the notice ran, 'Montmartre, capital of Paris, will be shaken by one of those events which sometimes change the face of the earth. The Cabaret du Chat Noir will leave the boulevard Rochechouart, which its presence has adorned for a long time, and establish itself on the rue de Laval.'[7]

The ironic tone and the eccentric humour became a permanent feature of popular entertainment at the Chat Noir and at many other *cabarets artistiques*. In 1885, after the Chat Noir had settled into its new quarters, Rodolphe Salis increased the variety of the programmes by installing a little theatre, or *theâtricule*, for the performance of short, comic plays with song and musical accompaniment. The first of these plays, entitled *Madame Garde-Tout*, featured a song in which overstatement of the obvious resulted in a parody of aphoristic sayings:[8]

> A staircase which had no stairs
> Would definitely not be a staircase:
> A staircase, it must have stairs,
> Without which, it is no longer a staircase.

[6] Quoted in Guy Erismann, *Histoire de la chanson* (Paris: Pierre Waleffe, 1967), p. 129.
[7] *Le Chat noir*, 9 May 1885 (Geneva: Slatkine Reprints, 1971).
[8] Quoted in Herbert, *La Chanson à Montmartre*, p. 160.

The straightforward, deadpan tone here and the absurd humour created both by excessive logic and by stripping an object of its defining characteristics (stairs from a staircase) provided an important source for the literary sketches that Satie began writing in 1912 and for the annotations accompanying his piano pieces. The textual description for one of Satie's *Enfantillages pittoresques* (1913), for instance, tells of a king who liked his staircase so much that he never stepped on it and decided to have it stuffed.

Salis asked the painter Henri Rivière to take charge of the new *théâtricule*, and before long Rivière introduced his shadow puppets. He suspended a sheet of oil paper in front of the theatre's small stage, plunged the room into darkness, and depicted a story contained in the verses of a popular cabaret song by moving cardboard silhouettes back and forth behind the paper. The device of the shadow puppet was so successful that cabaret artists began writing plays—sometimes with incidental music—for the new shadow theatre, and by 1886–7 shadow plays were a nightly event at the Chat Noir. Other cabarets featured the genre as well. A shadow play called *Noël*, with text by Vincent Hyspa and music by Erik Satie, was performed sometime in 1881 at the Auberge de Clou.[9]

From 1886 to 1897, when the Chat Noir closed, the programmes were extremely diverse. Poetry and song still formed the staple of cabaret entertainment, but in addition the Chat Noir now offered shadow plays in twenty or thirty tableaux with music and scenery, and humorous dramatic sketches called *pochades*.[10] *Poètes-chansonniers* generally performed their works during the interludes between the tableaux of a shadow play or the acts of a *pochade*. This order of events called attention to the variety of the repertoire.

Many late nineteenth- and early twentieth-century *cabarets artistiques* emulated the diverse offerings of the Chat Noir. An announcement in the entertainment column of *Le Gaulois* on 13 January 1903, describes the assortment of songs and satiric plays featured at another venue, the Cabaret des Arts: 'At the Cabaret des Arts, after having heard the singers present their works, nothing

---

[9] Herbert discusses the acclaim at the Chat Noir of a shadow play, *L'Épopée*, and writes that soon after the 1886 performance, shadow plays became a nightly event. Although he does not specify exactly when this occurred, we can presume that shadow plays could be seen every night at the Chat Noir beginning in 1886. See ibid., pp. 163–4. For information on *Noël*, see ibid., p. 294.

[10] One shadow play, *La Tentation de Sainte-Antoine*, bore the subtitle 'féerie à grand spectacle en 2 actes et 20 tableaux' (spectacular fantasy in 2 acts . . .). Such a subtitle suggests the elaborate scale of some of these plays. See Steven Moore Whiting, 'Erik Satie and Parisian Musical Entertainment 1888 to 1909', Master's thesis (Musicology), University of Illinois, Urbana, 1983, p. 45.

is funnier than to see them perform the *Game of the Car*, a comic-lyrical play which they wrote themselves. In this new adaptation of Faust, V. Hypsa plays Mephisto; Montoya, the Doctor Faust; J. Ferny, Valentin; and J. Varney, Bob Wagner.'[11] Similarly, at the Cabaret de la Roulotte, founded in 1896, the programme included performances of old French songs by professional actors, *saynetes* (short drawing-room comedies for two or three performers), dramas, and popular songs. The better-known Cabaret des Quat'z'arts, which opened in 1893, contained a café room decorated in a Gothic style, a restaurant above the café, and a special hall where *poètes-chansonniers* presented their songs, shadow plays, and *pochades*. Moreover, the Quat'z'Arts was the first cabaret to organize an outdoor event. This carnivalesque parade known as the *Vachalcade* took place in 1896. A horse-drawn caravan led the procession, tamed bears and horses followed, and the *poètes-chansonniers* brought up the rear. *Vachalcades* were still popular in the early 1920s when they were sponsored by the cabaret La Vache Enragée, along with racing and swimming events, bull-fights, smoking competitions, and ceremonies in honour of the newly elected Muse of Montmartre.[12]

The Chat Noir set a precedent for the artistic décor of later cabarets, the performance of songs and shadow plays, and the attraction of a clientele composed of intellectuals, aristocrats, and wealthy professionals. Indeed from its inception, the Chat Noir was a fashionable place for Parisians to go at night. Guy Erismann, a historian of French song, reports that all those regarded in Paris as 'snobs, moneyed people, and overfed financiers and politicians' attended the Chat Noir on Friday evening which became known as *jour chic*.[13] Members of the French aristocracy were first prompted to visit the cabaret when they learned that they would be in the company of such famous intellectuals as Victor Hugo, Giuseppe Garibaldi, Émile Zola, and Léon Daudet. The critical acclaim of the shadow plays, moreover, allured foreign aristocrats—the Emperor of Brazil, the kings of Portugal, Greece, and Serbia—as well as French princes, grand-dukes, generals, and diplomats. Cabaret entertainment continued to be fashionable among aristocrats well into the 1920s. The *poète-chansonnier* Henri Fursy, who ran his own cabaret artistique, was so highly esteemed that when a foreign

---

[11] 'Spectacles divers', *Le Gaulois*, 13 Jan. 1903, p. 3.
[12] The information in this paragraph is drawn from Herbert, *La Chanson à Montmartre*, pp. 315, 383, 414.
[13] Erismann, *Histoire de la chanson*, p. 129.

dignitary visited Paris, Fursy was asked to serve as the representative of French song.[14]

THE CAFÉ-CONCERT

A popular institution which catered to shopkeepers, clerks, and other members of the lower and middle classes, rather than to an élite clientele, was the café-concert.[15] Unlike the cabaret, the café-concert made no artistic claims; the tone of its repertoire was unabashedly coarse and down-to-earth. 'If the song of today is often stupid enough to make us cry,' wrote the critic Jules Lemaître of the banality of the Parisian café-concert song, 'it does not fool the world, it presents itself for what it is, it does not aspire to be literature.'[16] Cabaret songs such as the *chansons sociales* of the poet and singer Aristide Bruant were occasionally transported to the café-concert by singers like Yvette Guilbert who hoped to improve the quality of entertainment and introduce a more sophisticated brand of humour.[17] For the most part, however, the two institutions maintained a separate repertoire.

The café-concert was a much older establishment than the cabaret. Its origins can be traced to the *musicos* of the mid-eighteenth century, where people ate and drank while enjoying the entertainment of singers and fairground performers, and to the private singing societies known as *caveau* and *goguette*. During the 1840s, garden cafés similar to the musicos opened on the Champs-Élysées. Since entertainment at these cafés was strictly musical, they were called *cafés-chantants*. Parisians came to converse, to drink, and to hear singers dressed in evening clothes perform on the stage of a kiosk, while an orchestra provided accompaniment. By the time of the Second Empire (1852), new pavilions and more elaborate musical programmes had prompted a modification of the establishment's name to café-concert. Institutions modelled after the

---

[14] Herbert, *La Chanson à Montmartre*, pp. 163, 305.

[15] Theodore Zeldin's discussion of café-concert performers reveals much about the clientele, since entertainers and audience members generally came from the same background. See Zeldin, *France 1848–1945: Taste and Corruption* (Oxford: Oxford University Press, 1980), pp. 352–9.

[16] Quoted from Lemaître's *Impressions de Théâtre, 1888–1898* in François Caradec and Alain Weill, *Le Café-Concert* (Paris: Atelier Hachette/Massin, 1980), p. 152.

[17] See Bettina Knapp and Myra Chipman, *That was Yvette: The Biography of Yvette Guilbert, The Great Diseuse* (New York: Holt, Rinehart, and Winston, 1964), p. 57. Guy Erismann discusses the occasional appearance of *chansons artistiques* at the café-concert. See Erismann, *Histoire de la chanson*, p. 127.

outdoor café-concerts soon opened in the winter and employed artists from the Champs-Élysées.[18]

The winter café-concerts, like their summer counterparts, attracted a clientele accustomed to smoking a pipe and drinking beer while they listened to romances and light songs. Indeed the atmosphere at the café-concert was extremely informal. During the early years of its popularity, the seating arrangement resembled that of the regular café, with isolated tables and chairs grouped around each. By 1867, rows of chairs with trays attached at the back for drinks had replaced the separate café tables. The greater formality of this seating arrangement did not substantially alter the ambiance. The hall was still noisy and smoke-filled, and people still felt free to interject comments, join in the singing, and come and go as they liked.[19]

Unlike the cabaret, where patrons had to pay for their seats from the time that *théâtricules* were installed, the café-concert charged no admission. It would have been difficult to require an entrance fee because the café-concert programme was long and had no formal beginning. Rather it moved through rounds of song called *tours de chant*, with each series featuring a different singer; interludes of dance; and a section in which the singers presented an operetta, a short musical comedy, or a *saynete*.[20] Patrons came to hear a *tour de chant* or a musical comedy, and then returned home or went to another café-concert. In fact, the audience was discouraged from staying for the entire evening because drinks were obligatory, and one had to renew the order after each *tour de chant* or each act of an operetta.[21]

The unrefined ambiance of the café-concert and the earthy, often obscene humour which characterized many of the songs caused French critics to write disparagingly about the quality of entertainment. Such descriptions, if subjective, are valuable none the less for the picture they present of the café-concert milieu. Jules Bertaut, for example, associated the café-concert with the following sights and sounds:[22]

[18] For more information on the origins of the café-concert, see Guy Erismann, *Histoire de a chanson*, p. 112; France Vernillat and Jacques Charpentreau, *Dictionnaire de la chanson rançaise* (Paris: Larousse, 1968), p. 49; Caradec and Weill, *Le Café-Concert*, pp. 7–12.

[19] See Caradec and Weill, *Le Café-Concert*, p. 12, and Zeldin, *France 1848–1945*, pp. 351–2, for information on the atmosphere and seating arrangement of the café-concert.

[20] Georges Chepfer discusses the format of the café-concert programme in 'La Chansonette t la musique au café-concert', in Ladislas Rohozinski, ed., *Cinquante ans de musique rançaise de 1874 à 1925*, 2 vols. (Paris: Librairie de France, 1925), ii. 236.

[21] See Caradec and Weill, *Le Café-Concert*, p. 19.

[22] Jules Bertaut, *Les Belles Nuits de Paris* (Paris: Flammarion, 1927), p. 120.

The café-concert and its coarse music played by a vulgar orchestra, the café-concert and its unbreathable atmosphere of smoke clouds and stale smells of beer, the café-concert and its lewd or obscene refrains taken up in chorus by the public, . . . the café-concert and its musical comedy actresses and its mundane *diseurs* and its duet singers and its soldiers and its odors of gas, of oil, of cellar moisture, and of sweaty linen.

The *diseurs* and soldiers mentioned at the end of this passage represented two of many performance types, each specializing in a particular vocal genre or style of singing. The *diseur* performed in a declamatory style and used subtle vocal nuances and bodily gestures to emphasize certain words. The soldier, known among French audiences as the *comique troupier*, was naïve and bumbling and wore pants, vest, and cap which were too small. Another popular type, the *gommeuse*, was portrayed by lustful female singers who used their silly, often salacious love songs as vehicles for seducing the public and arranging rendezvous after the performance. Although patriotic songs (*chansons d'actualité*) also played an important role in café-concert entertainment, the comic genres were the most popular. Two widely performed comic genres were the *chansons idiotes et scabreux* or idiotic and licentious songs— commonly interpreted by the singer Dranem—and the *scie*, a monotonous, nonsensical refrain repeated over and over by the singer expressly to annoy people.[23]

The lack of artistic aspirations at the café-concert did not prevent a certain overlap between this popular establishment and more cultivated institutions like the opera and the theatre. During the Second Empire, opera singers and singers of *opéra-comique* frequently participated in operettas and even *tours de chant* at the café-concert, where they generally performed under a different name. In addition, the café-concert served as a training-ground for performers of musical theatre who wished to try out their songs in an informal atmosphere before auditioning at the Parisian *théâtres lyriques*.[24] Thus the transfer of such theatrical genres as *saynetes* and operettas to the café-concert was paralleled by an exchange of performers between the two institutions.

THE CIRCUS

Theatrical genres had an impact on cabaret and café-concert, and on a third establishment: the circus. In the late nineteenth and early

---

[23] Caradec and Weill, *Le Café-Concert*, pp. 24, 130, 152, and Erismann, *Histoire de la chanson*, p. 122.

[24] See Caradec and Weill, *Le Café-Concert*, pp. 20, 38.

twentieth centuries, the Nouveau Cirque, the Cirque Médrano, and the Cirque de Paris were the principal circuses in Paris. Although all three tended to feature a similar kind of entertainment and to compete for first running of new acts, the Nouveau Cirque and the Cirque Médrano were more popular than the Cirque de Paris and received greater coverage in the entertainment columns of Parisian dailies.[25] In June of 1900, the special attraction at the Nouveau Cirque was not an acrobatic stunt or a clown act but a musical play about American cowboys and Indians, entitled *Les Indiens Sioux*. The play was introduced by the new orchestral director, a man named Wittmann, who had formerly conducted the orchestra at the Opera balls. An advertisement appearing in *Le Figaro* on 3 June 1900 suggests that spectacle in *Les Indiens Sioux* took precedence over story-line: 'The cowboys of the Far-West interest one with their surprising lasso exercises, their curious songs, their very colourful dances, and their special way of mastering untamed horses.'[26] Along with *Les Indiens Sioux*, the programme at the Nouveau Cirque featured four equestrian performers known as the Frères Frediani—who formed a human column on top of a horse— and a comic *pochade* about tourists who get caught in a downpour.[27]

One of the special appeals of the Nouveau Cirque was an apparatus which transformed the track into a lake. This technical innovation enabled the circus to present aquatic events in which dancers on toe perched on the lake's waterlilies, clowns impersonating hunters plunged into the water in pursuit of a wild boar, and a host of other imaginative feats occurred.[28] The transformation of the track, a ten-minute procedure, was a spectacle in itself, and many Parisians came primarily to witness this event. First the carpet covering the track was rolled up and carried away on a special wagon, then the track was submerged, and finally sprays of water filled the newly created basin.[29]

In an effort to make the aquatic scenes which followed as

[25] A sample of announcements which appeared in the entertainment columns of two Parisian dailies, *Le Figaro* and *Le Gaulois*, between 1900 and 1920, reveals the extensive coverage of events at the Nouveau Cirque and the Cirque Médrano. Le Nouveau Cirque was particularly active in the early 1900s and the Cirque Médrano in the later 1910s (after around 1915).

[26] 'Spectacles et Concerts', *Le Figaro*, 3 June 1900, p. 4.

[27] These remaining contents of the programme are mentioned by Adrian in his *Histoire illustrée des cirques parisiens d'hier et d'aujourd'hui* (Paris: Adrian, 1957), p. 65.

[28] Cocteau describes the dancers on the waterlilies in his *Portraits-souvenir: 1900–1914*, edited by Pierre Georgel (Paris: Librairie Générale Française, 1977), p. 89. The hunting scenes culminating with a plunge into the water are recounted in the 'Spectacles et Concerts' column of *Le Figaro*, 7 May 1900, and 13 May 1900.

[29] Adrian, *Histoire illustrée des cirques*, p. 55.

interesting as the machinery, the Nouveau Cirque designed visually elaborate pantomimes in which the performers typically jumped into the water at the end. For instance, the year 1889 opened with a new pantomime called *La Noce à Chocolat*, featuring two of the most popular clowns at the Nouveau Cirque: the English exile Foottit and his black partner the Spanish Chocolat.[30] Like most theatrical events at the circus, *La Noce à Chocolat* was richer in spectacle than in plot. It wove a comical story about a groom whose wife is abducted on her wedding day, into a series of humorous pranks culminating with a dive into the water. Aquatic events at the Nouveau Cirque also took the form of *féeries* or fantasies such as *Une féerie à Seville* or *Carnaval de Venise* in which dances, songs, comic numbers, and exotic costumes evoked a foreign culture through Parisian eyes.[31]

Circuses in Paris prided themselves on the variety of their entertainment. At the Nouveau Cirque, theatrical genres ranging from *pochades*, pantomimes, and *féeries*, to musical plays and operettas were featured alongside clown and acrobatic acts, juggling performances, magic shows, spectacles with trained animals, and lion-taming numbers. The same diversity prevailed at the Cirque Médrano. On 9 October 1915, the *Figaro* announced the following events: 'At the Cirque Médrano. This evening, debut of the two Abravys, xylophonists. Every evening, we acclaim the Russian dancers, the Fratellini clowns, Dario and Ceratto and the athletes, and male and female equestrian performers.'[32] Individual acts could be as eclectic as the entire assemblage. One of the most popular clowns at the Cirque Médrano, who went by the name of Grock and exemplified a particular class of entertainer known as the musical eccentric, presented slapstick numbers, parodistic renditions of American marches and cakewalks on the clarinet, and more serious, virtuosic performances on a vast number of other musical instruments. Debussy's 'General Lavine—eccentric' (*Préludes*, Book II) clearly drew inspiration from this type of performer.[33] Likewise the popular transformists, who donned in quick succession the physiognomy of various celebrated artists and politicians, may have spawned Satie's frequent adoption of poses that caricatured particular human types (the moralist, the pedant).

---

[30] In Satie's *Menus propos enfantins* of 1913, the reference to 'chocolat' in 'Valse du chocolat aux amandes' may have been intended as a pun invoking the popular clown Chocolat as well as the candy.

[31] For further information on *La Noce à Chocolat* and on the *féeries* at the Nouveau Cirque, see Adrian, *Histoire illustrée des cirques parisiens*, p. 57.

[32] 'Spectacles et concerts', *Le Figaro*, 9 Oct. 1915, p. 4.

[33] For a description of Grock's act, see Georges Fréjaville, *Au music-hall* (Paris: Aux éditions du monde nouveau, 1923), p. 190.

The flexibility of the circus as a forum for many different kinds of entertainment prompted the *Association des Artistes Lyriques* to hold its annual festival at the Nouveau Cirque rather than at a café-concert or a *bal-restaurant*.[34] The festival, which took place on 14 June 1913, featured popular café-concert singers who took turns impersonating equestrians, acrobats, clowns, and other circus roles. The evening ended with a pantomime aptly entitled *Tout à l'eau*, in which everyone fell into the water.[35] Such reciprocity between popular establishments set an important precedent: by July and August of 1917, programmes at the Nouveau Cirque were offering a blend of circus and café-concert entertainment, with singers performing alongside clowns.

The crowd which attended the circus spectacles tended to be more diverse than either the cabaret or the café-concert audiences. It comprised a striking mixture of lower-class Parisians who stood in the large gallery overhanging the arena and paid two francs for their tickets, middle-class Parisians who sat in the rows of small rocking chairs below and paid three francs, and wealthy aristocrats who rented their *loges*.[36] The Nouveau Cirque, moreover, attracted foreign visitors interested in the latest marvels of the French capital. According to the entertainment writer for *Le Figaro* reporting on 18 May 1900, 'The Nouveau-Cirque has become the rendez-vous for all those who come to Paris to see the sights.'

THE FAIR

Whereas Parisian circuses provided year-round entertainment and concentrated on visual spectacle, the fair or *fête foraine* of the early twentieth century was an annual event which featured not only spectacle, but game, lottery, and the sale of food, household items, and decorative objects.[37] Circus and fair also differed in their

---

[34] A *bal-restaurant* was a restaurant with two or more stages for dancing and for song performances.

[35] Adrian, *Histoire illustrée des cirques*, p. 71.

[36] One should note that this kind of diversity was particularly characteristic of the audience at the Nouveau Cirque. Adrian, *Histoire illustrée des cirques*, pp. 53–4. Another writer on the circus and the music-hall, Pierre Bost, observes that the audiences at other Parisian circuses generally belonged to the lower and middle classes. Elegant, wealthy spectators appeared more frequently at the Nouveau Cirque and at Parisian music-halls. See Bost, *Le Cirque et le music-hall* (Paris: René Hilsum, 1931), pp. 42–4.

[37] In his *Forains d'hier et d'aujourd'hui*, Jacques Garnier describes the *fête foraine* as an annual event awaited with great excitement in Parisian towns and cities. However since Milhaud mentions visiting the fair each Saturday evening, perhaps fairs in Paris and outlying towns took place more frequently. Unless otherwise specified, information on the fair is drawn from Garnier, *Forains d'hier et d'aujourd'hui* (Orléans: Les Presses, 1968), pp. 9–18, 122–3, 274–86.

setting. By the late nineteenth century, the Parisian circus was housed permanently in a vast indoor amphitheatre. The fair, on the other hand, took place outdoors with merchants displaying their goods in booths along an avenue, and entertainers presenting their shows in stalls or temporary theatres. Visitors strolled down the boulevards and stopped whenever something caught their eye. Thus they might pause to examine the pottery and china at one booth, purchase a piece of lace or silk from another vendor, and then enjoy a waffle or a pastry in the garden café of another stall while listening to military music performed by the local garrison.

The games, lotteries, shooting galleries, and exhibition booths were grouped together on a separate boulevard. One of the most popular games was the *jeu de massacre*. The player purchased five cloth balls, and used them to try and knock down at least five marionettes representing a bride, a groom, and assorted members of a wedding party. The booth-keeper advertised his game with cries of 'Aim well at the groom! Hit the mother-in-law!' In 1921, Cocteau wove the *jeu de massacre* into a scene of his *pièce-ballet, Les Mariés de la tour Eiffel*, in which the child of the bride and groom tossed bullets and 'massacred the wedding party'.

At the lotteries people played the roulette wheel, or bet on ducks and pigeons which had been assigned numbers. Meanwhile shots could always be heard from the galleries where players were firing at little dots on a cardboard box, at strings from which bottles and packages of tobacco were suspended, or at balls moving on top of a tiny jet of water. In a special kind of gallery called the *tir à surprise*, or surprise shooting gallery, one tried to hit targets adorning a series of little cupboards. When the player succeeded, the doors of the cupboard opened and out popped tiny dolls who danced or performed a simple scene to the accompaniment of a music-box. Such a game may have inspired the title of Satie's *Jack in the Box*, written for a pantomime by Jules Dépaquit.

Along the avenue, mixed in with the lotteries and shooting galleries, were the exhibition booths. For a small entrance fee, fair visitors could choose between fortune-tellers, freak shows, animal tamers presenting their wild beasts, and magicians who pulled rabbits out of a hat, made flowers appear in their buttonholes, and produced freshly laid eggs from thin air. The fairground theatres offered more spectacular conjuring tricks. One of the most popular was the decapitation act in which the performer cut off his head and presented it on a plate to his baffled spectators. This act—which evokes the decapitated head of John the Baptist delivered to Princess Salome on a silver platter—underscores the enormous appeal of the Salome legend for artists (Aubrey Beardsley, Gustav Klimt, and Richard Strauss) and for the Parisian public during the early decades of the twentieth century. The decapitation theme at the fairground also reveals the remarkable interplay between 'art' and everyday entertainment in Paris. Among Satie's circle, the Salome story was invoked in Cocteau's scenario for Milhaud's ballet *Le Bœuf sur le toit*. Towards the end of the ballet, the Barman decapitates the Policeman with a big fan, whereupon the Red-headed Woman performs a dance with the Policeman's head, 'ending up standing on her hands like the Salome in Rouen cathedral'.[38]

An event which rivalled the decapitation act in popularity was the one-man show by a performer such as Frego, who had successfully mastered the art of conjuring, juggling, mimicking, and ventriloquy, and could execute all of these roles in quick succession, as well as perform in his own sketches and comedies. Dramas and *féeries* were also frequently performed. The *féeries* usually contained twenty or thirty tableaux with mimed scenes, narration, and electric lighting. Often one half of a theatre's programme was devoted to a *féerie*, and the other half to clown, acrobatic, magic, and animal-taming numbers.

Many of the fair spectacles presented at show booths and theatres were identical to circus performances.[39] In addition, the *fête foraine* had its own circus which was housed in a tent on a prominent part of the fairground and featured the standard fare of clowns, acrobats, equestrians, and animal tamers. Each afternoon several clowns, some dancers and acrobats, and a group of small animals including ponies and monkeys advertised the evening programme at the circus by executing a few brief scenes on a small platform called the *parade*. The performance, also known as the *parade*, was accompanied by a brassy orchestra, and presided over by a master of ceremonies who cried intermittently, 'Stop the music! Ladies and Gentlemen, come near. Tonight there will be only a single grand evening of festival.' He went on to list each of the programme's attractions, marvel at the modest prices of the seats, and urge spectators to hurry and purchase their tickets before it was too late. The circus performers 'en parade' echoed the manager's cries by brandishing placards marked with the prices of the cheapest seats. Then the orchestra struck up a boisterous march and some of the

---

[38] This quote is part of Milhaud's narration of the scenario in *Notes without Music*, p. 87.

[39] Georges Fréjaville discusses the impact of fair attractions—lion tamers, magicians, ventriloquists, transformists, freaks—on the programmes of the Nouveau Cirque and the Cirque Médrano. See Fréjaville, *Au music-hall*, p. 246.

spectators approached the ticket counter while others wandered off to another fair booth which may also have featured a *parade*. Such preview entertainment was a pervasive feature at the fairground and became the theme of Satie's celebrated ballet of 1917.

The mixture of merchant booths, circus acts, and theatre shows in an outdoor setting made the fair a real festival enticing people from every social class. According to Georges Fréjaville, there was such a hullabaloo that the *fête foraine* often resembled a formidable factory, where the sounds of orchestras and of the brass were practically covered by the steam whistles, the rumbling of the merry-go-round, the explosion of shots, the grating of turnstiles, the merchant calls, and the cries of the crowd.[40] The presence of these sounds reproduced, in microcosm, the simultaneity of urban life in Paris. Darius Milhaud was so impressed that he wrote his ballet *Le Bœuf sur le toit* in a polytonal language intended to imitate the fairground's jangling simultaneity.

THE MUSIC-HALL

The diversity of entertainment at the fairground found its match in the eclectic offerings of cabaret, café-concert, and circus. A fifth popular establishment, the Parisian music-hall, exceeded cabaret, café-concert, circus, and fair in the range of its offerings and in its tendency to absorb genres from other milieux. Jacques Feschotte, a French historian, calls the music-hall a vast pot-pourri where, 'following its many influences and aspirations, one accumulates the most diverse elements, ranging from the better to the worse, from the basest realism to the most extreme poetry'.[41]

The flexibility and diversity of music-hall entertainment makes an historical account difficult. Yet as numerous as the ingredients of music-hall programmes were, there was a principal source and spawning ground for this entertainment in the café-concert. Indeed the music-hall grew out of the café-concert, and for a time in the late nineteenth- and early twentieth-centuries the two terms and two kinds of establishments coexisted with only the subtlest of differences between them. Music-hall ultimately replaced café-concert, but the shift was a gradual one first prompted by the appearance of theatrical entertainment at the café-concert in 1867.

Prior to 1867, strict regulations forbade the use of costumes and the presentation of plays, dances, pantomimes, and operettas at the

[40] Fréjaville, *Au music-hall*, p. 246.
[41] Jacques Feschotte, *Histoire du music-hall* (Paris: Presses Universitaires de France, 1965), p. 6.

café-concert. The *tour de chant* was the attraction, and it was performed by singers in evening dress or city clothes. Such rules were enforced by theatre officials who feared competition from popular establishments. In 1867, however, the director of the popular café-concert called the Eldorado engaged an actress from the Théâtre-Français, Mlle Cornélie, to recite passages from Racine and Corneille, including 'Le Songe d'Athalie' and 'Imprécations de Camille'. Although Mlle Cornélie wore an evening gown rather than a costume, she was the first tragedian to perform the classic repertoire on the stage of a café-concert. The public was enthusiastic, and the Parisian press immediately launched a campaign demanding an end to costume censorship in the café-concerts. Theatre officials soon responded to the pressure. On 31 March 1867, Camille Doucet, the managing director of Parisian theatres, authorized the presentation of actors in costume and the inclusion of short plays, pantomimes, operettas, and dance and acrobatic numbers, all with scenery changes, on the programme of the café-concert.[42]

The lifting of theatrical censorship marked an important step in the transition from café-concert to music-hall. A history of the music-hall by Jacques-Charles goes so far as to state that in 1867 with the appearance of costumes, plays, and dances at the café-concert, the music-hall was born. Similarly, Jacques Feschotte calls the combination of songs, operettas, ballets, and clown acts at the Alcazar d'Hiver in 1867, a 'music-hall formula'.[43] Both writers imply that when theatre, dance, and circus attractions appeared beside songs, the café-concert soon became a music-hall.[44]

The term 'music-hall' was not actually applied to a Parisian popular establishment, however, until 1893 when Joseph Oller opened the Olympia.[45] Furthermore, in newspaper entertainment columns from the turn of the century, reports about the conversion of a café-concert into a music-hall indicate that café-concerts still existed at this time. The question then arises: if, after 1867, both café-concert and music-hall were popular institutions featuring a

[42] Censorship at the café-concert and the birth of the music-hall are discussed by Jacques-Charles in *Cent ans de music-hall: histoire générale du music-hall de ses origines à nos jours* (Geneva: Éditions Jeheber, 1956), pp. 100–1, and by Caradec and Weill in *Le Café-Concert*, p. 39. Jacques-Charles was the author of several revues and the director of some music-halls during the late 1910s and early 1920s.

[43] Jacques Feschotte, *Histoire du music-hall*, p. 84.

[44] Guy Erismann likewise argues that when scenery and costumes were permitted at the café-concert, and 'song became a spectacle', the music-hall emerged. Erismann, *Histoire de la chanson*, p. 122.

[45] The designation of the Olympia as a 'music-hall' is discussed by Caradec and Weill in *Le Café-Concert*, p. 183.

programme of songs, dance and acrobatic interludes, and operettas, precisely how did the two differ? A brief article in the 18 September 1899 issue of *Le Figaro*, announcing the transformation of the popular café-concert, the Casino de Paris, into a music-hall, reveals both the fine line between café-concert and music-hall and the distinguishing features of each.[46] The author of the article, Abel Mercklein, emphasized the dramatic physical change of the Casino de Paris. What had formerly been an old café room, he reported, was now a 'veritable palace, a marvel of comfort and good taste'. Gold leaf covered both wall and ceiling, and new *loges* had been installed to accommodate a larger, wealthier clientele. One gathers from this brief description that music-hall and café-concert differed principally in the greater spaciousness and opulence of the former. With its 'palatial' setting and its decorated interior, the music-hall looked much like an elegant theatre.

Historical accounts reveal additional ways in which the environment of the music-hall resembled that of a theatre. Seats were arranged in rows facing the stage, and there were no trays or tables for drinks. In fact, people were not permitted to smoke or drink during the performance, and the minimum drink requirement was replaced by an entrance fee.[47] Such changes in physical setting made the music-hall more formal than the café-concert.

Yet formality was largely the by-product of a design inspired by practical concerns. Music-halls needed grander auditoriums in order to accommodate the circus spectacles. At first, such acts were presented at the café-concert, but growing popular acclaim prompted directors to move their establishments to larger halls better suited to the acrobat, clown, and magic shows. Exactly when such halls became known as 'music-halls' seems to have varied with the establishment,[48] but the increasing importance of circus entertainment went hand-in-hand with the expansion and the transformation of café-concert into music-hall. Among the entertainers shared by circus and music-hall from approximately 1900 on, Jacques Feschotte cites clowns, gymnasts, rope dancers, stunt men, contortionists, magicians, conjurors, mime artists, classical dancers, exotic dancers, freaks, and animal trainers.[49] An announcement of the programme at the Casino de Paris, which appeared in *Le Figaro*

---

[46] See 'Spectacles et Concerts', *Le Figaro*, 18 Sept. 1899, p. 5.

[47] For information on the physical set-up of the music-hall, see Caradec and Weill, *Le Café-Concert*, p. 185.

[48] For instance, the Ba-Ta-Clan was built in response to the need for larger halls, but it was not known as a music-hall until 1913. See Jacques Feschotte, *Histoire du music-hall*, pp. 28, 89.

[49] Feschotte, *Histoire du music-hall*, p. 53.

on 4 July 1900, reveals the extensive borrowing of circus numbers at individual Parisian music-halls. The programme featured:[50]

Matsui, the Japanese Imperial Troupe; the de Charmion quartette performing acrobatic dances; Juliano Bozza and his musical clowns; the des Mullini, virtuosos on the cornet; and de Hormann, tight-rope walker and juggler.

The foreign names included in this announcement suggest an additional difference between café-concert and music-hall. Whereas the café-concert was unable to feature many circus acts, particularly those with elaborate technical apparatus, the music-hall included such entertainment regularly and, in the process, acquired an internationalism which the café-concert could only try and emulate. Circus and fair entertainers often came from Japan, Italy, Spain, and the United States and brought with them an air of the exotic. For example, the transformist Fregoli and the juggler Rastelli were both Italian performers who appeared frequently at Parisian fairs, circuses, and music-halls. Little Tich, a British comedian, and Loie Fuller, an American dancer, were two of the most celebrated music-hall stars of the late nineteenth and early twentieth centuries.[51]

A genre which epitomized the difference in scale and breadth between music-hall and café-concert was the revue. Both institutions featured revues, but only the music-hall revue—with its elaborate scenery, its electric lighting effects, its spectacular tableaux, and its mixture of fair, circus, theatre, dance, and song—became known as the *revue à grand spectacle*. According to Jacques Feschotte, the *revue à grand spectacle* commanded attention as 'the most original spectacle of the music-hall'.[52]

The revue originated in eighteenth-century Paris at the fair, and consisted of a succession of scenes in which the principal political, social, and artistic events of the year were 'passed in review'.[53] Two narrators, a *compère* and his leading lady, the *commère*, linked the diverse scenes together. The device of *commère* and *compère* was well-known to Jean Cocteau who introduced two human phonographs to narrate the events of his *pièce-ballet*, *Les Mariés de la*

---

[50] See 'Spectacles et Concerts', *Le Figaro*, 4 July 1900, p. 4.

[51] For more information on the internationalism of music-hall entertainment, see Feschotte, *Histoire du music-hall*, pp. 22, 69, 75. For information on Little Tich, see Jacques-Charles, 'Naissance du music-hall', in *Les Œuvres libres* 304 (Nov. 1952): 109.

[52] Feschotte, *Histoire du music-hall*, p. 76.

[53] For this information on the original meaning of 'revue' and on the early history of this genre, see Feschotte, *Histoire du music-hall*, p. 76, and Robert Dreyfus, *Petite histoire de la revue de fin d'année* (Paris: Fasquelle, 1909), pp. 2–3.

tour Eiffel, in 1921. In the late nineteenth century when café-concerts began presenting revues, they adhered to the concept of scenes based on *l'actualité*, or current events, and employed the *compère* and the *commère* as narrators. The music-hall revue, by contrast, reduced the portion about everyday events until it was merely a point of departure for a series of dances, comic and dramatic sketches, songs, circus numbers, and visually dazzling tableaux which evoked a tropical forest or an antique vision, or reproduced a well-known painting. Such a show became known as a *revue à grand spectacle*.[54]

Among the most opulent of the *revues à grand spectacle* were those held at the Folies-Bergère.[55] On 1 January 1903, the entertainment column of *Le Gaulois* provided a lengthy description of the latest such revue, entitled *La Revue des Folies-Bergère*:[56]

The Revue des Folies-Bergère includes no less than fourteen tableaux by Menessier . . . in the midst of which four hundred costumes designed by Gerbault emerge.

And everything marches by without the slightest hitch: the delicious album of postcards, with its charming *tableaux vivants*; . . . the admirable tableau of the Venice festival, perfect restitution of the sixteenth century . . .; the shimmering tableau of the Quai aux Fleurs; the exquisite and murmuring ensemble of Parisian petits-pierrots. . . .

I guarantee you that the eyes have something to feast upon.

And all of these splendours are interspersed with comic scenes.

The author of the column, Adrien Vely, went on to describe some of the comic scenes. Two sketches, which featured the actor Maurel in the role of police commissioner, fulfilled the original purpose of the revue in their humorous treatment of current social problems: security in the streets of Paris, and the expulsion of fraternities. Another performer entertained the audience with magic acts, songs, dances, and pantomimes. Larger spectacles consisted of a special comic effect in which a sidewalk and boulevard moved forward while buses rolled in place, and a performance by a company of musketeers who executed military manœuvres. Throughout the performance, the orchestra was a crucial element. It linked the diverse tableaux and acts, and it played throughout, except when yielding place to a group of foreign musicians.

[54] Jacques Feschotte distinguishes between the café-concert's revue and the music-hall's *revue à grand spectacle* more clearly than any other writer. See *Histoire du music-hall*, p. 77.

[55] In 1886 the Folies-Bergère led the way in presenting the first *revue à grand spectacle*. See Caradec and Weill, *Le Café-Concert*, p. 78.

[56] 'Spectacles divers', *Le Gaulois*, 1 Jan. 1903, p. 4. *Tableaux vivants* appeared in *Les Mariés de la tour Eiffel*.

The impressive opulence and the intricacy of *La Revue des Folies-Bergère* were not necessarily typical of music-hall revues in the 1890s, 1900s, and early 1910s. Moreover, revues were not always the main staple of entertainment at café-concerts and music-halls. On a given evening in 1899 or 1900 some music-halls and café-concerts offered revues, several others presented an evening of songs performed by both popular vedettes and singing troupes, while still others selected certain elements of the *revue à grand spectacle* and presented these as independent attractions rather than scenes in a large theatrical show.[57]

Equally characteristic of music-hall programmes at the turn of the century was a division of the evening into a series of 'parties', one reserved for a play, one for dances and circus attractions, one for a short revue rather than a *revue à spectacle*, and one for song. For instance, the programme at the Scala (a music-hall) on 21 April 1903 was described by the author of the daily entertainment column of *Le Gaulois* as:[58]

the prototype of a varied programme likely to satisfy all publics without ever annoying: the *partie comique* with *Folles Têtes*, the extremely funny play in which Sulbac is incredible, along with Morton, a new arrival of an astonishing fantasy; the *partie élégante* with the exquisite Liane de Pougy in the revue *Viens foufoule* in which the animated tableaux are new artistic plays . . .; the *partie concert* with Paulette Darty, more than ever the public's idol, and Mayol, the fine *diseur*. There is here something which will satisfy the most difficult to please.

When such well-known stars as Paulette Darty and Mayol were featured, the *partie concert* was commonly reserved for last. Thus the *tour de chant* became the culminating point of the evening, with each singer performing as many as ten songs. For the 'Spectacle-Concert' which Cocteau organized in 1920, he placed the *tour de chant* right before the intermission. He also divided his evening into a series of 'parties' analogous to those he had witnessed at the music-hall.[59]

During the 1890s, as the revue became a popular form of café-concert and music-hall entertainment, it also entered the *cabaret*

---

[57] For example, on Tuesday, 4 July 1899, the daily entertainment column of *Le Figaro* announced the following events: at the café-concert Parisiana, a revue entitled *Paris sans facteurs*; at the café-concert Les Ambassadeurs, a group of singers including Yvette Guilbert and Sulbac; at the music-hall Olympia, a 'spectacle varié' featuring Little Tich, a 'grand ballet' called *La Fée des poupées*, and some jugglers, and at the Marigny Theatre, a 'ballet-féerie' entitled *La Fontaine des fées*. The 'spectacle-varié', the 'grand ballet', and the 'ballet féerie' were short theatrical forms which could also have appeared within the framework of a revue.

[58] 'Spectacle divers', *Le Gaulois*, 21 Apr. 1903, p. 4.

[59] For a detailed discussion of Cocteau's 'Spectacle-Concert', see Chapter 6.

artistique and the circus. The cabaret's and circus's adoption of the revue, like the music-hall's incorporation of *tours de chant* from café-concert and cabaret, reveals once again the easy flow between places of popular entertainment in Paris. In 1894, the Quat'z'Arts was the first *cabaret artistique* to present a revue. The title of the show, *Tout pour les Quat'Czars* (Everything for the Four Czars), contained a pun on the words 'czar' and 'art', which recalled the satiric humour and the word-games of cabaret songs. *Tout pour les Quat'Czars* had none of the sumptuousness of the *revue à grand spectacle*, nor did it claim to be anything more than a brief, simple, workshop farce performed by the *poètes-chansonniers* of the cabaret.[60] Soon, however, the influence of the music-hall became more potent. In 1895, the Treteau de Tabarin presented professional actors, rather than chansonniers and amateur performers, in a new *revue d'actualité* entitled *Paris sur le pont* (Paris on the Bridge). Gradually other *cabarets artistiques* began engaging choreographers, and costume and set designers, in order to produce a revue much like the *revue à grand spectacle*. Parades were added, special singers were lauded as the stars or *vedettes* of the evening, and dancing girls and acrobats appeared.[61] The spirit of satire still remained, but now it existed in a new context of spectacle.

Spectacle never overpowered satire, however. Indeed the penchant for parody, and the attempt to appeal to a more intellectual public kept cabaret revues distinct from their music-hall counterparts. On 9 February 1903, Nicolet, a writer for *Le Gaulois*, made reference both to the satiric humour and to the cabaret's continued interest in attracting a more selective clientele.[62] The humorous tone of his announcement seemed to mirror the irony and hyperbole of cabaret entertainment:

At the *cabaret artistique* Eugenie Buffet, one is not content to have the most select spectators, but one must also present plays in which the heroes are chosen from among the greatest names in France!

Par filons!, the fantasie-revue . . . which triumphed for the first time on Friday evening, involves a Duke of Lauxun, a Count of Maurepas, etc., etc.

Just as the revue was smoothly incorporated into a cabaret programme of shadow plays, *pochades*, and songs, so it joined the

---

[60] Michel Herbert discusses this early cabaret revue in *La Chanson à Montmartre*, p. 31.

[61] Georges Fréjaville discusses how the cabaret revues began to approach the scale of *revues à grand spectacle*. See *Au music-hall*, pp. 44–5.

[62] 'Spectacle divers', *Le Gaulois*, 9 Feb. 1903, p. 4.

other theatrical attractions—plays, pantomimes, *féeries*, *pochades* —at the circus. *Paris au galop*, a 'revue équestre et nautique', appeared at the Nouveau Cirque in 1889. Since this was the year of the Universal Exposition and the year which witnessed completion of the Eiffel Tower, the revue paid tribute to the new monument while adapting it to a popular setting. Each scene took place at the base of an Eiffel Tower constructed out of rope; equestrian performers galloped along the circus track to and from the Tower and—judging from the description of the revue as 'nautique'— dived into the lake at the end. The Eiffel Tower, a symbol of Paris and of modernism, later became the setting of Cocteau's 'pièce-ballet' *Les Mariés de la tour Eiffel* and of the last portion of the René Clair film, *Cinéma*, presented during the entr'acte of Satie's ballet *Relâche*.

Another revue, *À la cravache*, appearing at the Nouveau Cirque in November of 1890, featured a female equestrian performance and a parody of Cleopatra by the clown Foottit. Since Sarah Bernhardt was currently performing the role of Cleopatra, Foottit's parody may have included playful mimicry of her.[63] The content of both these spectacles suggests that at the Nouveau Cirque, revues did not differ all that much from the pantomimes and plays in which hunting and equestrian scenes culminated with a dive into the water.

Although cabaret and circus overlapped with café-concert and music-hall, the former two establishments also maintained their independence. One sign of this was the persistence of references in entertainment columns to certain popular establishments as 'cabarets'. Likewise circuses were not confused with other institutions, in part because they rarely incorporated song, and in part because the name circus generally appeared in the institution's title. The appellations 'café-concert' and 'music-hall', by contrast, were scarcely ever used in newspaper columns to identify individual entertainment spots, and the few times that the terms did appear the particular usage reflected the ambiguous distinction between the two establishments. For instance, in the late nineteenth century the appearance of the word 'concert' in the title of an institution— Eden-Concert, Concert Européen, Concert Mayol—indicated that the establishment was a café-concert; in the early twentieth century, the use of the word 'concert' persisted even when entertainment had expanded to include an elaborate revue with tableaux, songs, and circus numbers. Thus in 1915 at the Concert Mayol, *revues à grand*

[63] For a discussion of these revues, see Adrian, *Histoire illustrée des cirques parisiens*, p. 59.

spectacle were frequently advertised and stars like Mistinguett—who was described by *Le Figaro* as 'la grande étoile du music-hall'—presented dances and songs.[64]

It became increasingly difficult to distinguish between café-concert and music-hall during the post-war years as these establishments merged and the appellation 'music-hall' became permanent. According to two authorities on the café-concert, François Caradec and Alain Weill, the internationalism of music-hall entertainment was largely responsible for the demise of the café-concert. Caradec and Weill cite a principal event which, along with the frequent appearance of circus entertainers at the music-hall, ushered in the new internationalism: in 1917 the Casino de Paris featured a *revue à grand spectacle* entitled *Laisse-les tomber* in which the husband–wife team, Harry Pilcer and Gaby Deslys, performed syncopated dances to the accompaniment of a ragtime band. Once American rhythms had been incorporated into the Parisian revue, Caradec and Weill assert, the music-hall came into its own.[65]

The views of Caradec and Weill on the dividing line between café-concert and music-hall are closely corroborated by Jacques Feschotte, who states that the music-hall flourished between 1900 and 1914 but did not reach its apogee until after 1918.[66] Similarly, Jacques-Charles calls 1918 the end of the 'grande époque du café-concert' and the beginning of the 'grande époque du music-hall'.[67] Jacques-Charles selects 1918 because in this year the Casino de Paris presented another major revue called *Pa-ri-ki-ri*, featuring the couple Mistinguett and Maurice Chevalier in their debut performance. Mistinguett had begun to establish a reputation with her brutal athletic dance, the *valse chaloupée* or waltz with a rolling gait, which she presented at the Moulin Rouge and other music-halls. In *Pa-ri-ki-ri* she danced while Chevalier sang. The two were a tremendous success, and during the 1920s they dominated the music-hall scene. In the public eye they were not only stars, but popular heroes who epitomized the ideal of the romantic couple.[68]

During the closing years of the war, music-hall stars became more widely acclaimed and publicized, and the revues which had been popular since the turn of the century grew grander and more visually dazzling. Yet visual elements never overpowered the music. The

---

[64] See 'Spectacles et concerts', *Le Figaro*, 7 Aug. 1917, p. 4.
[65] See Caradec and Weill, *Le Café-Concert*, pp. 183–5. Jacques-Charles discusses the ragtime performance by Pilcer and Deslys in *Cent ans de music-hall*, p. 186.
[66] Jacques Feschotte, *Histoire du music-hall*, p. 29.
[67] Jacques-Charles, *Cent ans de music-hall*, p. 191.
[68] Theodore Zeldin describes the dreams and ambitions that Mistinguett and Chevalier represented for the Parisian public. See Zeldin, *France 1848–1945*, p. 359.

introduction of American syncopated dances often with an American band accompaniment assured the place of dance and music in the revue. Song also retained the prominent position which it had enjoyed at the café-concert. The *revue à grand spectacle* depended for its success, moreover, on the support of a music-hall orchestra which provided opening music, created links between the diverse elements of the show, performed accompaniments for most numbers, and offered a conclusion.[69] The role of the orchestral conductor was of utmost importance. He supervised the collaboration between authors and choreographers of a revue and, as a composer, he provided the musical décor; this involved arranging or adapting popular songs, refrains, and dance tunes, and in each case selecting music which suited the choreography and the scenery.

THE CINEMA

As early as 1898, the music-hall included film on its list of attractions. In this way, it served as the 'cradle' of an institution which ultimately superseded it in popularity: the cinema.[70] The origins of film can be traced to the Edison Kinetoscope of 1894, which preserved photographs on strips of film and placed the film on a revolving wheel.[71] Only one person could view the pictures at a time, by looking into a peep-show box. Before long the Edison Kinetoscope was taken to France where it provided quite an inspiration for two brothers, Louis and Auguste Lumière. In February 1895, they patented a device called the Cinématographe which could photograph and project films. Although they first regarded their device as a useful scientific invention, the Lumière brothers soon realized its potential as a new form of popular entertainment. On 28 December 1895 they exhibited the Cinématographe before the Parisian public in the basement of the Grand Café on the Boulevard des Capucines. This room had recently been transformed into a divan called the Salon Indien which the Lumières rented out. They arranged the café chairs in rows before the screen and charged a modest admission of one franc per twenty-minute show.[72]

[69] Jacques Feschotte discusses the prominent position of song and of the music-hall orchestra in *Histoire du music-hall*, pp. 28, 56, 121.

[70] Fréjaville calls the music-hall the 'cradle of the cinema' in *Au music-hall*, p. 276.

[71] For a discussion of the Edison Kinetoscope as an important forerunner of the film projector, see George C. Pratt, *Spellbound in Darkness: A History of the Silent Film*, rev. edn. (Greenwich, Conn.: New York Graphic Society, Ltd., 1973), p. 7.

[72] Both the invention of the Cinématographe and public reaction to this new invention are discussed in David Robinson, *The History of World Cinema* (New York: Stein and Day, 1981), pp. 1, 17, 18. Cocteau may have attended the show at the Salon Indien.

The display of the Cinématographe marked the first significant exploration of film technique by the French. Strangely enough, Parisian newspapers made no mention of the event. The public, however, responded with enthusiasm, and by January audiences were lining up to see the Lumière exhibit. The early films belonged to a genre which Georges Fréjaville calls 'documentary'.[73] They consisted of everyday events and bore such titles as *High Seas at Brighton, Arrival of a Mail Train, Workers Leaving a Factory*, and *Men Playing Cards*. In addition there were scenes based on reality but a bit more out of the ordinary, for example: *Demolition of a Wall* and *The Turn-out of the Leeds Fire Brigade*. The use of everyday subject-matter in film may well have provided an impetus for Cocteau's choice of the fairground setting in *Parade*, the American bar setting in Milhaud's ballet *Le Bœuf sur le toit*, and the fashionable resort in Milhaud's *Le Train bleu*. In the 1920s, moreover, when film began to play an important role in ballet, opera, and theatre, Satie's final ballet *Relâche* included a film in which the action shifted abruptly from a Paris rooftop to a theatre, a chess game, and a funeral procession.

From the Salon Indien, film projections moved to the Parisian café-concert and music-hall. In 1898 the Olympia and the Folies-Bergère were the first music-halls to present the Lumière Cinématographe, and other music-halls and café-concerts (the Eldorado and the Ba-Ta-Clan) soon followed. In 1897, moreover, fairground entertainers throughout Europe exhibited film as an independent show. Crowds flocked to the annual fairs in Paris and in small towns to witness the photographing and projecting of street scenes. The fairground cinemas, at first small modest theatres, became larger and were furnished with electric lights and steam organs.[74]

Beginning in the early 1900s in Paris, animated pictures could thus be seen at music-halls and café-concerts, at fairs, and even at circuses which often concluded their programmes with a few film projections. In addition there were special theatres like the Grands Magasins Dufayel and the Musée Grevin which devoted their evening programmes to film.[75] The subject-matter of these films revealed a subtle shift in emphasis from the everyday event to the superhuman venture, as well as an interest in scenes which had been

[73] See Fréjaville, *Au music-hall*, p. 276.

[74] Robinson discusses the emergence of film shows at European fairgrounds in *The History of World Cinema*, p. 25.

[75] The Musée Grevin may have been a special theatre devoted to magic shows; in his *Histoire du music-hall*, Jacques Feschotte writes, 'The Musée Grevin continues to present good magic spectacles' (p. 75).

choreographed on a theatre stage first and then filmed. For example on 9 March 1903, *Le Gaulois* announced a film programme at the Musée Grevin, consisting of several ambitious explorations of nature, followed by a choreographed dance: 'At the Musée Grevin, every day a large public comes to the Cinématographe to applaud the Ascension of the peak of Mont Blanc, the Exploration of the forest of Canada, and the curious cakewalk dance.'[76]

Other announcements of the period mentioned the orchestral accompaniment which seems to have been an important feature of film shows. An especially detailed announcement, appearing in *Le Figaro* on 1 July 1902, described the military music and the drum fanfares accompanying a film of the Franco-Russian and Spanish celebrations which was shown at the Grands Magasins Dufayel.[77]

During the war years France came to rely increasingly on American films, and the American serial became one of the most popular forms of entertainment. French film-goers, Jean Cocteau among them, had an opportunity to see two of the most famous Hollywood serials, *The Perils of Pauline* (1914) and *The Exploits of Elaine* (1915).[78] The adventures of Pauline provided an important source of inspiration for Cocteau's conception of the Little American Girl in *Parade*.

American comic films became popular in Paris during the later years of the war. Milhaud and Blaise Cendrars joined Cocteau in attending the American comedies, particularly those of Charlie Chaplin which began to be screened in Paris in around 1917. Chaplin's films appeared at Parisian cinemas and movie-halls such as the Salle Marivaux and the Pathé-Palace, which had been built after the establishment of permanent cinemas in 1905 and 1906.[79] A preview announcement of one of these films, offered in *Le Figaro* on 2 January 1920, suggests that French audiences were especially impressed by Chaplin's malleability: the ease with which he placed himself in so many different kinds of situations, and evoked laughter each time:[80]

---

[76] 'Spectacles divers', *Le Gaulois*, 9 Mar. 1903, p. 4.

[77] See 'Spectacles et Concerts', *Le Figaro*, 1 July 1902, p. 3.

[78] *The Perils of Pauline*, with the famous actress Pearl White in the title role, focused on the adventures of the lovers, Pauline and Harry. In one episode, Pauline was captured by gypsies and Harry set off in search of her. In another episode, the lovers were threatened by some escaped lions. See Pratt, *Spellbound in Darkness*, pp. 135–6.

[79] For instance on 2 Dec. 1919, the film *Charlot va dans le monde* was playing at the Salle Marivaux. See 'Spectacles et Concerts', *Le Figaro*, 2 Dec. 1919, p. 3. On 9 January 1920, *Charlot s'évade* appeared at the Pathé-Palace. 'Spectacles et Concerts', *Le Figaro*, 9 Jan. 1920, p. 4.

[80] 'Spectacles et Concerts', *Le Figaro*, 2 Jan. 1920, p. 4.

Charlie can make a film [in which he] becomes a soldier, leads the life of a dog, jumps, dances, skates: Charlie spreads laughter. Have you seen him when he's sick? He is irresistible. You can judge for yourself by going to the Pathé Palace to see . . . *Charlie takes a cure* . . . the most astounding and beautiful show of the unrivalled artist.

Occasionally Chaplin also performed in person, in sketches with the English pantomime troupe known as Fred Karno. He and the Karno troupe enjoyed a series of successes at the Folies-Bergère where Chaplin presented magic tricks and played the part of the drunkard. Other music-halls also featured shows by the brilliant comic artist. On 25 September 1917, *Le Figaro* announced a performance by 'M. Charlot, the king of comics, in a sketch with his troupe' at the Casino de Paris.[81] Milhaud's delight with Chaplin is reflected in his composition in 1919 of the score of *Le Bœuf sur le toit*, originally entitled *Cinéma fantasie* and intended to be an accompaniment for a Chaplin silent film.

For a time, music-hall and cinema enjoyed a close collaboration. Such performers as Chaplin, Maurice Chevalier, and Max Linder moved easily from one realm to the other, and film was frequently the subject of music-hall sketches when it was not actually part of the evening's attractions.[82] Yet in the mid-1920s, as the popularity of film increased, halls devoted entirely to film flourished at the expense of music-halls. By 1935, cinema usurped the reputation which the Parisian music-hall had enjoyed since the turn of the century.[83]

The popular haunts of Cocteau, Satie, Milhaud, Poulenc, and Auric extended beyond cabaret, café-concert, circus, fair, music-hall, and cinema to *bals-musettes* (dancing halls), *guinguettes* (suburban cabarets), and ordinary bars. Yet the six establishments discussed in the preceding pages represented the favourite and most widely frequented milieux. Entertainment was not exclusively French, as we have noted; at circus, music-hall, and cinema, American popular dance and music likewise played a role. The American genres also spilled over into public squares and outdoor arenas, hence furthering the fluid interchange which characterized the motley world of Parisian popular entertainment.

---

[81] 'Spectacles et Concerts', *Le Figaro*, 25 Sept. 1917, p. 4. Jacques-Charles discusses Chaplin's work with the Fred Karno troupe in 'Naissance du music-hall', pp. 135–6.

[82] Gustave Fréjaville describes how the cinema furnished subject matter for two music-hall shows in 1919: the first, *Faisons du Ciné*, was presented at the Alhambra; the second, *Les Coulisses du Cinéma*, appeared at the Olympia. See Fréjaville, *Au music-hall*, pp. 276–9.

[83] On the demise of the music-hall, see Caradec and Weill, *Le Café-Concert*, p. 185, and Feschotte, *Histoire du music-hall*, p. 29.

2

The Arrival of American Popular Music and Dance on the Parisian Scene

A FASCINATION with American popular culture animated Paris during the war years and after. Music-hall revues featured perform-ances of syncopated dancing, and movie-palaces offered screenings of the latest Charlie Chaplin films. In a light-hearted vignette published in *Vanity Fair* in August 1922, Ezra Pound poked fun at the pervasive influence of America on the French. He observed that, with the Parisian craze for American culture reaching its peak, 'any simple national American custom such, let us say, as the roasting of marshmallows, might easily become here a profession'.[1]

Six months earlier, the American critic Edmund Wilson had offered his comments on the lure of recent American inventions:[2]

Young Americans going lately to Paris in the hope of drinking culture at its source have been startled to find young Frenchmen looking longingly toward America.

In France they discover that the very things they have come abroad to get away from—the machines, the advertisements, the elevators, and the jazz—have begun to fascinate the French at the expense of their own amenities. From the other side of the ocean the skyscrapers seem exotic, and the movies look like the record of a rich and heroic world full of new kinds of laughter and excitement.

Wilson also identified a desire among French people, and particu-larly among artists, to obliterate the memory of the war, which they associated with a decadent, old-world sensibility. For the French, Wilson explained, America's vitality and freshness offered a welcome alternative to European conventions. In the eyes of French artists, moreover, the bright spontaneity which they found so appealing was encapsulated in American popular culture. Thus they looked to American film, dance, and music as new sources of creative inspiration.

[1] Ezra Pound, 'On the Swings and Roundabouts: The Intellectual Somersaults of the Parisian vs. the Londoner's Effort to Keep his Stuffed Figures Standing', *Vanity Fair* 18 (Aug. 1922): 49.
[2] Edmund Wilson, 'The Aesthetic Upheaval in France: The Influence of Jazz in Paris and Americanization of French Literature and Art', *Vanity Fair* 17 (Feb. 1922): 49.

AMERICAN ENTERTAINERS

Pound and Wilson were correct in suggesting that the Parisian interest in American culture reached its peak after World War I. Yet American popular entertainers had been an important presence in Paris since the 1880s. Many Americans performed at the circus and music-hall, two popular centres for foreign entertainment. Others introduced their music and dance in cafés, theatres, and public squares. Often these events received wide press coverage and attracted enormous crowds. It is likely that the artists Satie, Cocteau, Milhaud, Poulenc, and Auric were members of the crowds, since they shared a receptivity to a wide range of popular entertainment. As the oldest of the five, Satie may have attended American performances as early as the end of the 1870s when he became a teenager. Cocteau and Milhaud, born in 1889 and 1892 respectively, may have been present during the first decade of the twentieth century, and Poulenc and Auric probably attended by about 1912.[3]

The first major concert of American popular music to take place in late nineteenth-century Paris featured military band music, performed by Patrick Gilmore and his 22nd Regiment Band during the Universal Exposition of 1878. Between 3 and 4 July, Gilmore's band gave thirteen concerts, including three at the Trocadero.[4] The Trocadero programme featured opera overtures by Rossini and Weber, the Andante from Beethoven's Symphony No. 5, the 'Star Spangled Banner' sung by the famous opera singer Miss Lilian Norton (known by 1879 under the name Lillian Nordica), and a fantasy on American country airs.[5]

French audiences were captivated by the 22nd Regiment Band, which was at that time the finest on the American scene, and Gilmore received a medal from the French government. The success of his concerts encouraged an exchange of repertoire between French and American military bands. One indication of this was the appearance in an American arrangement of the famous march *Le Père la Victoire* by the French composer Louis Ganne. Above the title, the sheet music cover contained the following advertisement:

[3] This speculation about the years in which Satie, Cocteau, Milhaud, Auric, and Poulenc first attended performances of American music and dance in Paris is based on the assumption that each was old enough to attend such events by about the age of 12 or 13.

[4] For information on the dates of Gilmore's performances, see Marwood Darlington, *Irish Orpheus: The Life of Patrick S. Gilmore, Bandmaster Extraordinary* (Philadelphia: Olivier-Manoy-Klein, 1950), p. 93.

[5] The entire programme was outlined in a brief article in *Le Figaro* which reviewed Gilmore's concerts at the Trocadero. 'Petit Courrier de l'Exposition', *Le Figaro*, 4 July 1878, p. 1.

'THE FAMOUS SUCCESS as played by Gilmore's band'. Such acclaim marked America's rise to prominence as a leader in the band movement and set the stage for the second European tour by an American band.[6]

In spring of 1900, John Philip Sousa brought his civilian band to Paris for the Universal Exposition. From 6 May to 4 July the band entertained Parisians with afternoon concerts at the Invalides.[7] A column in *Le Figaro* on 10 May 1900 offered a brief account of the warm reception which Sousa's band had received the day before:

An enormous crowd, beginning at 3:30 p.m. on the promenade of the Invalides, where John Philip Sousa and his orchestra gave their daily concert.

Not an empty chair around the kiosk where the American flags fluttered; and great success as on all the days.

The programmes presented by Sousa's band were more diverse than those of Gilmore, ranging from transcriptions of classical compositions, to fantasies, celebrated marches by Sousa—*The Washington Post*, *King Cotton*, and *The Stars and Stripes Forever*— and pieces composed by other members of the band.[8] From the Parisians' point of view the most exciting part of the tour and, indeed, the event without precedence in France, was the performance of American syncopated dance music. The band introduced such popular American dance pieces as Kerry Mills' *At a Georgia Campmeeting*, and Abe Holzmann's *Smoky Mokes*, *Hunky Dory*, and *Bunch o' Blackberries*, all in Sousa's own band arrangements.[9]

Parisian newspapers of the period consistently referred to these pieces as 'cakewalks', perhaps because Sousa himself used this

[6] Charles Hamm offers a short account of Gilmore's career. See *Music in the New World* (New York: W. W. Norton, 1983), pp. 289–90. For statements on the success of Gilmore's European tour, see Marwood Darlington, *Irish Orpheus*, p. 93; and Carolyn Bryant, *And the Band Played On* (Washington, DC: Smithsonian Institution Press, 1975), p. 28.

[7] There is some discrepancy over the dates of Sousa's performances in Paris. *Le Figaro* announced concerts from 9 May to 15 May 1900, and then again from 3 to 20 July. Sousa's autobiography cites band performances from 6 May to 4 July. See John Philip Sousa, *Marching Along: Recollections of Men, Women, and Music* (Boston: Hale, Cushman & Flint, 1928), pp. 182, 197.

[8] Sousa's programmes were listed in the 'Exposition' column which appeared every day in *Le Figaro*. Classical compositions (performed in transcriptions for band) included excerpts from *Tannhaüser*, Liszt's *First Hungarian Rhapsody*, the finale to the third act of Puccini's *Manon*, and Strauss's *The Blue Danube*. Dalby's *My Old Kentucky Home* and a collocation called *Songs from the Cotton Pickers Way Down South* (based on Stephen Foster tunes) illustrate the fantasies performed, and Arthur Pryor's *The Blue Bells* (for trombone) was one of several works composed by band members.

[9] The dance pieces which Sousa introduced are cited in Rudi Blesh and Harriet Janis, *They all Played Ragtime* (New York: Oak Publications, 1966), p. 74.

term.[10] American popular-song writers of the late 1890s, on the other hand, called the same pieces 'ragtime hits'.[11] Such discrepancies in terminology indicate that the words 'cakewalk' and 'ragtime' were not used with much precision in this period. Accordingly, it is difficult to provide separate definitions of each or to argue that the label 'cakewalk' was more accurate for *At a Georgia Campmeeting* than 'ragtime'. Since cakewalk preceded ragtime, however, it is possible to identify early differences in the meanings of the two terms.

The cakewalk originated as a plantation dance in which slave couples kicked and strutted in a parody of their masters' manners. The masters gathered around to watch the fun and, missing the point of the joke, awarded a cake to the dancers who 'did the proudest movement'.[12] In 1877 the American musical comedy team, Harrigan and Hart, brought the cakewalk to the vaudeville stage where it appeared in a song and dance number called *Walking for dat Cake*. In 1883, moreover, black performers (as opposed to whites in blackface) danced a piece entitled *Sam Johnson's Cakewalk* in Harrigan and Hart's *Cordelia's Aspirations*. By 1898 the cakewalk had established its name in the world of American musical theatre: *Clorindy*, or, *The Origin of the Cakewalk*, with music by Will Marion Cook, lyrics by Paul Laurence Dunbar, and a song and dance performance by the black comedian, Ernest Hogan, was presented on Broadway for a white audience; and *In Dahomey*, also by Cook and Dunbar and featuring cakewalk stepping by the black comedy duo, Bert Williams and George Walker, became an important Broadway success. In 1903 *In Dahomey* moved to London where it played for seven months and inspired an international cakewalk craze. At the same time, cakewalk contests were being organized in ballrooms all over the United States.[13]

The music which accompanied the cakewalk was originally an unsyncopated form of march music. According to Edward Berlin,

---

[10] *Le Gaulois* contained several articles announcing Sousa's performances of 'cakewalks' at the Nouveau Théâtre. 'Spectacles divers', *Le Gaulois*, 19 Apr. 1903; 'Spectacles divers', *Le Gaulois*, 21 Apr. 1903.

[11] In 1900, the popular-song writer R. M. Stults referred to dance pieces by Kerry Mills as 'rag-time hits'. According to Edward Berlin, Stults's judgements are representative of the views held by contemporary songwriters. See Berlin, *Ragtime: A Musical and Cultural History* (Berkeley: University of California Press, 1980), pp. 81, 104.

[12] Rudi Blesh quotes this description of the cakewalk by Shephard N. Edmonds, a prominent Negro entertainer. See Blesh and Janis, *They All Played Ragtime*, p. 96. For a brief statement on the plantation origins of the cakewalk, see Berlin, *Ragtime*, p. 104.

[13] For historical background on the cakewalk, see Blesh and Janis, *They All Played Ragtime*, p. 96; Marshall Stearns, *Jazz Dance: The Story of American Dance* (New York: Macmillan, 1968), pp. 117–18; Charles Hamm, *Music in the New World*, p. 342.

syncopated rhythms began to appear in cakewalk marches as the popularity of the dance increased and reached a peak. This use of syncopation coincided with the early growth of ragtime and may have reflected ragtime influences. During the late 1890s and early 1900s new syncopated figures, including patterns based on small rhythmic divisions of the beat, were introduced in ragtime and then incorporated in cakewalk music as well. Eventually the two forms of dance music merged. Americans may have continued to dance the cakewalk after 1904, but by that year ragtime had become the general term for American syncopated dance music.[14]

A dance piece like *At a Georgia Campmeeting* can therefore be considered a syncopated cakewalk or an early rag. However, since 'cakewalk' was the term which gained currency in Paris in the early 1900s, and since the dance contests which became a Parisian craze were always associated in the United States with the cakewalk, this term seems most appropriate for a discussion of American syncopated dance music in Paris between 1900 and 1903.

Sousa's performances of the cakewalk made a powerful impression on Parisian audiences: the dance became a rage almost overnight in Paris salons. A German magazine from 1900 contained a photograph bearing the caption 'Der Cake Walk in einem pariser Salon', and in the 10 June 1900 issue of the *San Francisco Chronicle* the music for one strain of *Bunch o' Black-berries* appeared above an article entitled 'Paris Has Gone Rag Time Wild'.[15] Within two years of Sousa's tour, popular institutions in Paris were featuring the cakewalk as a principal attraction in their spectacles. On 24 October 1902, for example, the Nouveau Cirque made use of its special machinery which transformed track into lake, in order to present *Joyeux Nègres*, a show described by *Le Figaro* as a 'grande pantomime américaine nautique, avec le Cake-voalk [*sic*]'.[16] The clowns Foottit and Chocolat were two of the prominent performers, and the show's entertainment comprised an exhilarating billiard game, a boxing match, bathers, songs, and 'a burlesque dance, the cakewalk, which is unlike anything we know and which people have demanded to see again'. *Joyeux Nègres* was so successful that its first run was extended for nearly a year. In the 1 January 1903 issue of *Le Gaulois*, a review of the 100th performance reported that at the end

---

[14] Edward Berlin identifies early differences between cakewalk and ragtime, explains that the term 'cakewalk' referred principally to a dance and only briefly to the music accompanying this dance, and describes how cakewalk music and ragtime eventually became one and the same. See Berlin, *Ragtime*, pp. 13–14, 104–6.

[15] Blesh and Janis, *They All Played Ragtime*, p. 81. Photographs of these press clippings appear opposite p. 81.

[16] 'Spectacles et Concerts', *Le Figaro*, 24 Oct. 1902, p. 4.

of the show members of the audience leapt out of their seats and danced the cakewalk late into the night.[17]

The Nouveau Cirque not only made the cakewalk part of its 'grande pantomime', but also began holding cakewalk contests in which any amateur could participate. The first took place on 21 March 1903 and its success prompted a second contest, advertised by *Le Gaulois* on 3 April and scheduled for the end of the month:[18]

At the Nouveau Cirque, the second cakewalk contest . . . promises to be as brilliant as the first. A considerable number of interesting prizes, some of great value, have been collected by M. Houcke for the contest, which will have this particularity in common with the competition of 21 March, namely that professionals will not be admitted, and it will be reserved for amateurs alone. The number of couples who participate cannot exceed 25; therefore it is wise to sign up now . . .

The year 1903 was rich in performances of American popular dances in Paris, for it also marked the return of Sousa and his band. Between 19 April and 1 May, Sousa conducted marches, cakewalks, and arrangements of classical compositions at the Nouveau-Théâtre. A review in *Le Gaulois* spoke glowingly of the first concert, pronounced Sousa an 'artiste', and urged anyone who believed in honouring 'la Musique' to attend the band's performances:[19]

Is it necessary to speak to you of Sousa, this extraordinary orchestral conductor, this 'march king', whose already famous works, even more than his stay in Paris during the Exposition of 1900, have made him popular among us? Go hear him and see him at the Nouveau-Théâtre, where he is giving a short series of concerts twice a day; go applaud the 'Washington Post', the Cake-Walk, and the 'Stars and Stripes Forever', so many pieces which are for him the object of indescribable ovations.

The efforts of Gilmore and particularly of Sousa to export American popular music to Paris were reinforced by the activities of French music-hall stars. During the early 1900s popular French singers and dancers were frequently engaged to perform in American vaudeville shows and on Broadway. When they came to the United States, the French entertainers learned American songs and dances, and often they brought this repertoire back with them

---

[17] 'Spectacles divers—soirée parisienne', *Le Gaulois*, 1 Jan. 1903, p. 4. As late as February 1905, the entertainment column of *Le Figaro* announced a revival of *Joyeux Nègres* and welcomed the idea, 'because no one has been able to forget the triumphant success of this passionate dance and everyone wants to see it again'. See 'Spectacles et Concerts', *Le Figaro*, 25 Feb. 1905, p. 5.

[18] 'Spectacle divers', *Le Gaulois*, 3 Apr. 1903, p. 3.

[19] 'Spectacles divers—soirée parisienne', *Le Gaulois*, 21 Apr. 1903, p. 3.

to Paris.[20] The singer and dancer Anna Held, to name one, was a performer in Parisian music-halls before she married the American producer Florenz Ziegfeld and came to the United States in 1900.[21] Between 1901 and 1908 Ziegfeld mounted many shows for Held in New York. In 1907 he created the first of his yearly spectaculars, the Ziegfeld Follies. In accordance with Anna Held's suggestion, Ziegfeld modelled his Follies after the Parisian music-hall revue. French entertainers were often invited to perform, and it seems probable that they, like Held, returned to Paris and introduced an assortment of American Tin Pan Alley songs and dance tunes.[22]

Ziegfeld's interest in foreign talent was shared by the Shuberts, a family of New York producers. In early January 1920, the Shubert Brothers opened an office in Paris in hopes of becoming better acquainted with French performers. Announcements appearing in *Variety Magazine* soon after the office opened indicate that the Shuberts were actively searching for foreign singers and dancers to supplement their American talent. On 9 January 1920, for example, *Variety* reported that the popular French music-hall entertainer, Mistinguett, had accepted a contract with the Shuberts and was coming to New York in March with her partner Maurice Chevalier. Another headline announced that Irving Berlin would soon arrive in Paris to arrange music for Mistinguett's show.[23] At the end of January, moreover, a headline on the front page of *Variety* ran: 'Shuberts scouring abroad for musical comedy talent . . . American

[20] I owe this information on the avenues of musical exchange between America and France to Russell Sanjek, former Vice-President for Public Relations of Broadcast Music Inc. Information was obtained from correspondence, 4 December 1983 and during a telephone conversation on 15 December 1983. Charles Stein explains that vaudeville consisted of the following acts: animal act; comedy skit; juggler; magician; one or two acrobats; pantomimist; dancing couple; singer; one-act play. Typically each vaudeville number featured a different performer. See Charles Stein, ed., *American Vaudeville as Seen by its Contemporaries* (New York: Alfred Knopf, 1984), p. 10.

[21] Anna Held claimed to be a native of Paris but, according to Gerald Bordman, she was actually born in Warsaw. She performed in Dutch and British music-halls, as well as Parisian nightspots, until Ziegfeld first spotted her in a London performance at the Palace. See Bordman, *American Musical Theatre: A Chronicle* (New York: Oxford University Press, 1978), p. 147. The year 1900 for Held's arrival in the United States is an approximation.

[22] Tin Pan Alley was the name for the popular-song industry centred around 28th St. in New York, beginning in the late 1890s, and for the style of song brought out by publishers working in this part of town. For a lengthy discussion of Tin Pan Alley publishers, song-writers and song styles, see Charles Hamm, *Yesterdays: Popular Song in America* (New York: W. W. Norton, 1979), pp. 284–390.

[23] For the announcements on Mistinguett and on Irving Berlin, see *Variety*, 9 Jan. 1920, p. 4. It is not certain that Irving Berlin actually wrote the music for a show starring Mistinguett. Gerald Bordman does not mention such a show. Nevertheless, the Shuberts' interest in arranging a collaboration between Irving Berlin and Mistinguett is noteworthy, because it reveals the reciprocity of performers between France and the United States, and the possible presence of American songwriters in Paris.

Vaudeville List Thought To Be Exhausted For Needs of Shuberts' Musical Shows. Have Over 150 Native Acts Under Contract'. One can assume, then, that the Shubert Brothers and Florenz Ziegfeld were responsible for bringing an increasing number of foreign singers, dancers, and possibly song-writers to the American stage.

The Parisian music-hall entertainer Gaby Deslys was one of several performers who responded to the American interest in foreign talent. Deslys made her American debut on 27 September 1911 in *The Revue of Revues*, a vaudeville show presented by the Shuberts at New York's Winter Garden theatre.[24] She played the part of a young Parisian orphan who aspires to stardom and performed a long dance-skit as part of her act. On 20 November 1911, the Shuberts featured Gaby Deslys in another musical at the Winter Garden entitled *Vera Violetta*. This show marked the first of Deslys's several successful appearances with the American stage performer, Harry Pilcer. The couple stopped the show every night when they danced and sang 'The Gaby Glide', with music by Louis Hirsch and words by Pilcer. In 1913, in a musical farce entitled *The Honeymoon Express* which opened on 6 February at the Winter Garden, Deslys donned a bespangled gown and danced with Pilcer; beginning on 25 December 1915, the couple performed ballroom dances and syncopated stepping in Irving Berlin's revue *Stop! Look! Listen!*[25]

Deslys married Harry Pilcer soon after her participation in Berlin's revue and remained active in the United States as a performer on Broadway and in vaudeville until 1917. On 12 October 1917, *Variety* announced that she had withdrawn from a show called *Suzette*, at the Globe theatre in New York, in order to accept a starring role in a new revue at the Casino de Paris.[26] Upon her return to Paris, Deslys arranged for Pilcer to perform syncopated dances with her in the pioneering revue *Laisse-les tomber* which, according to Milhaud and other French musicians and historians, introduced Parisians to 'jazz'.[27] Since Deslys and Pilcer had first performed and then perfected their syncopated dances in New

[24] For the information on Deslys's performances which appears in this paragraph, see Gerald Bordman, *American Musical Theatre*, pp. 271, 273, 286, 312. One additional example of a Broadway musical starring Deslys was *The Belle from Bond Street*. The show was imported from France and presented in New York beginning on 30 March 1914. See Bordman, p. 294.

[25] The syncopated stepping was performed to ragtime songs of Irving Berlin. Bordman notes that these songs often opened with rhymed dialogue. See Bordman, *American Musical Theatre*, p. 312.

[26] *Variety*, 12 Oct. 1917, p. 4.

[27] Milhaud discusses the arrival of a jazz band at the Casino de Paris in 'Les Ressources nouvelles de la musique', *Esprit nouveau* 25 (1924): n.p. According to Jacques-Charles,

York, their dance in *Laisse-les tomber* probably bore a closer resemblance to the American stage performances of syncopated dancing than any other cakewalk or ragtime dance which Parisians had seen. The couple stepped to the accompaniment of a seven-piece American band called the Jazz Kings led by the black drummer Louis Mitchell.[28]

Two years after Deslys had completed her engagement at the Casino de Paris, she returned to New York. On 24 October 1919 a column in *Variety* announced that she would head a new revue at the Century Grove theatre, and noted that she would also 'be allowed to fill in any other theatrical engagement she desires'.[29] Deslys returned to Paris later that year and, in all likelihood, introduced Parisians to the dances and songs she had performed in New York. Her theatrical activities were announced regularly in *Variety* until her death in 1920. Such close attention on the part of an American entertainment weekly indicates the extent of Deslys's reputation in the United States.

The activities of Harry Pilcer were also covered regularly in *Variety*.[30] In November 1920, the magazine announced Pilcer's appearance with an American troupe, the Jackson Girls, in the revue *Paris qui jazz* at the Casino de Paris. On 3 June 1921, *Variety* reported the closing of an outdoor Parisian dancing resort directed by Pilcer—perhaps a *bal-musette* or dancing hall where he had been teaching American syncopated dances.

Many other American entertainers brought their own repertoires to Paris. Typically these performers were vaudeville actors and dancers who could easily transfer their talents to the Parisian music-hall, *bal-musette*, or small theatre. A glance at the 'In Paris' column of *Variety* during the 1910s indicates how frequently Americans appeared in France. The entertainers Doris Thayer, Betty Washington, and Jimmy Fletcher, and the dancers Maurice and Florence Walters are just a few of the names mentioned.[31] Elsie Janis, a

Deslys and Pilcer brought the first jazz performance to France. See *Cent ans de music-hall: histoire générale du music-hall de ses origines à nos jours* (Geneva: Éditions Jeheber, 1956), p. 186.

[28] See below for more information on the Jazz Kings. Robert Goffin deduces that Louis Mitchell's band provided the accompaniment for the revue at the Casino de Paris. See Goffin, *Jazz: From the Congo to the Metropolitan* (New York: Da Capo Press, 1975), pp. 68–9.

[29] *Variety Magazine*, 24 Oct. 1919, p. 4.

[30] See the 26 Nov. 1920 (p. 2) and 3 June 1921 (p. 2) issues of *Variety*. Jacques-Charles discusses the enormous success of Harry Pilcer in Paris. See *Cent ans de music-hall*, p. 187.

[31] *Variety Magazine*, 21 Dec. 1917, p. 4; 14 Feb. 1919, p. 4; 14 Nov. 1919, p. 4. Doris Thayer, formerly with the Ziegfeld Follies, was performing in Paris at the Palais de Glace. Betty Washington performed at the Olympia and the Folies-Bergères. Jimmy Fletcher appeared with several American troupes at the Alhambra music-hall. Maurice and Florence Walters were featured in the revue *Pa-Ri-Ki-Ri* at the Casino de Paris.

popular American vaudeville entertainer, was one of the most active. In 1917, just after the United States declared war on Germany, Janis came to Paris and performed a variety act for a short time. Then she left for the front line in France where she entertained the troops with a vaudeville act comprised of sketches and imitations of American stars singing the patriotic songs that were flooding the United States. Janis imitated Sarah Bernhardt, Ethel Barrymore, Will Rogers, Eddie Foy, and many others.[32] Back in Paris in spring or summer of 1918, she gave a gala performance at the Théâtre des Champs-Élysées for 4,000 soldiers. Among the songs she performed were George M. Cohan's *Over There* and Shelton Brooks's *The Darktown Strutters' Ball*. After two years in London, Janis returned to Paris in May 1921 to appear in *La Revue d'Elsie Janis*, a show created especially for her at the well-known music-hall, the Apollo. Since she spoke French fluently, Janis sang American songs in her own French translations.[33]

Paving the way for Elsie Janis's acclaim during the war years and early 1920s were the remarkable dance performances of the husband–wife team of Vernon and Irene Castle. The Castles, and the black bandleader James Reese Europe who became their musical director, played a more important role than any other American performers in popularizing American syncopated dance music in Paris during the second decade of the century. They first came to France in the fall of 1911 when Vernon Castle, a dancer and actor, was engaged by the French music-hall director Jacques-Charles to perform in a revue at the Olympia music-hall.[34] Vernon agreed to participate if Irene were hired as well, and the French consented. The inclusion of Irene enabled the Castles to captivate Parisian audiences with the latest American social dances.

A dancing craze had erupted in the United States at approximately the time that the Castles left for Paris. The popular white social dance known as the two-step was superseded by a group of vernacular dances from the West and the South which began penetrating the higher echelons of white society. These energetic, 'belly-rubbing' dances bore titles like Turkey Trot, Grizzly Bear, Monkey Glide, Chicken Scratch, and Bunny Hug, and called for a

---

[32] Janis, *So Far, So Good!* (New York: E. P. Dutton & Co., Inc., 1932), pp. 181, 185, 204, 217.

[33] Janis's return to Paris was announced in the 20 May 1921 issue of *Variety*.

[34] Jacques-Charles explains that when he became director of the Olympia music-hall in 1911, he visited the United States in search of American talent. There he discovered and engaged Vernon Castle. See *Cent ans de music-hall*, p. 176. Irene Castle discusses Vernon's encounter with Jacques-Charles in Irene Castle with Bob and Wanda Duncan, *Castles in the Air* (Garden City, NY: Doubleday, 1958), p. 50.

fast, syncopated music with its basis in ragtime.[35] Tin Pan Alley songwriters, always alert to changing tastes, responded to the craze by incorporating the syncopated rhythms of the new dances into their songs. The ragtime-inspired songs were then able to function as dance pieces whether they were sung or played.[36]

During the Castles' first few months in Paris, they learned of America's new passion for syncopated dancing from newspaper clippings sent from home by Irene Castle's mother. Excited by the prospect of introducing Parisians to American dances, Vernon and Irene decided to create a unique finale for their revue at the Olympia in which they performed some of the new steps. They developed their own version of the Grizzly Bear and the Texas Tommy, and rehearsed it to the tune of Irving Berlin's *Alexander's Ragtime Band*. In her memoirs, Irene Castle describes their first dance performance when the revue opened in March 1912:[37]

If the American version was rough, ours was even rougher, full of so many acrobatic variations that I was in the air much more often than I was on the ground. The French audience was enthusiastic. They stomped their feet and clapped their hands and yelled 'Bravo'. They stood up at the end of the number and cried out 'greezly bahr' until we appeared again.

After performing in the revue for two weeks, the Castles left the Olympia. They received their next engagement sometime in late April or early May of 1912 when Louis Barraya, the proprietor of a Parisian supper club called the Café de Paris, booked them as dancers. The café had a small string orchestra, and one can speculate that Vernon and Irene Castle presented the musicians with a repertory of Tin Pan Alley tunes which were then performed as dance accompaniments. A favourite ragtime tune of the Castles', *Too Much Mustard*, seems to have become especially popular among the French, for it was performed and recorded by numerous French gypsy orchestras during the 1910s.[38]

---

[35] The two-step was a late nineteenth-century dance that developed out of the galop. It was a fast dance, performed to music in 2/4 or 6/8 time. See Charles Hamm, *Music in the New World*, p. 305. Further information on the new syncopated dances which replaced the two-step in Hamm, *Music in the New World*, p. 400; and Marshall Stearns, *Jazz Dance*, p. 96.

[36] Stearns and Hamm describe the response of Tin Pan Alley songwriters to the new dance craze. See Stearns, *Jazz Dance*, p. 95; and Hamm, *Music in the New World*, p. 365. Hamm notes that in the 1910s, syncopated dance rhythms became even more pervasive in popular songs. He also discusses the alliance of social dancing and popular song. See *Yesterdays*, p. 379.

[37] I. Castle, *Castles in the Air*, pp. 54–5.

[38] Information on French recordings of *Too Much Mustard* obtained during a conversation with Jean-Paul Guiter, head of the jazz department at RCA in Paris, on 18 June 1984. Two other ragtime tunes which Vernon and Irene Castle enjoyed dancing to were *Snookey Ookums* and *Everybody's Doing it Now*. See I. Castle, *Castles in the Air*, pp. 82, 85; and I. Castle, 'Things We Remember', *Liberty Magazine*, 31 Dec. 1927, p. 14.

Louis Barraya's Café de Paris attracted a clientele of wealthy Argentinians, Russians, and French nobility and thus was a much more elegant establishment than the Olympia. The Castles clearly preferred this aristocratic milieu; rather than returning to the music-hall, they divided their evenings between soirées and Barraya's café. Irene Castle recalls, 'As our reputation spread, we found ourselves greatly in demand for soirées and often did three a night . . . before we appeared for our midnight performance at the Café de Paris.'[39]

Sometime around May 1912, the Castles returned to the United States. They made two more trips to France, however, the first in the summer of 1913 and the second the following summer. On both occasions they danced at the Café de Paris and at an outdoor dancing resort in Deauville. By the spring of 1913 the Castles had invented a dance called the Castle Walk, consisting of a lift up on the beat rather than a step down, and it is likely that they introduced this dance to their Parisian fans in 1913. During their trip in 1914, the Castles probably performed the foxtrot, a syncopated dance which reflected their growing preference for an elegant ballroom manner.[40] The Castles created the foxtrot in collaboration with their famous musical director, James Reese Europe.

Vernon and Irene Castle first became acquainted with Jim Europe in the early 1910s. At this time he was directing black ensembles in performances of syncopated dance music at the Clef Club in New York. The Castles were aware that the syncopated music accompanying their lively dancing was played most skilfully by black musicians; accordingly, in 1913 they hired Europe to direct a dance band for their stage acts and for their studio teaching in New York.[41] Europe formed his Society Orchestra and began performing widely with the Castles. On 27 April 1914 he embarked with them on a 'Whirlwind Tour' of the United States. According to Europe's own account, it was during private sessions on tour that he played W. C. Handy's dance tune, *Memphis Blues*, which inspired the Castles' creation of the foxtrot. Soon after the tour, the Castles began dancing the foxtrot in public, and the dance held its

---

[39] I. Castle, *Castles in the Air*, p. 58.

[40] Both Charles Hamm and Marshall Stearns discuss the Castles' interest in an elegant, respectable style of social dancing. See Hamm, *Music in the New World*, p. 401, and Stearns, *Jazz Dance*, p. 97.

[41] The name of the New York studio where the Castles taught social dancing was the Castle House. Information on Jim Europe's role as music director for the Castles in Hamm, *Music in the New World*, pp. 401–3 and Robert Kimball and William Bolcom, *Reminiscing with Sissle and Blake* (New York: The Viking Press, 1972), p. 62. Among the dances introduced at the New York studio were the maxixe and the tango; in all likelihood, the Castles brought these with them to Paris where the tango had been fashionable since the turn of the century.

popularity long after other dances of the period had disappeared.[42]

Although the Castles may have engaged Europe as musical director before their second trip to Paris in 1913, they never brought Europe's Society Orchestra with them to France. Parisians did have an opportunity, however, to hear American syncopated dance music performed under Jim Europe's direction during World War I. In April 1917, when the United States entered the war, Europe enlisted in the 15th New York Regiment. The commanding officer of that regiment, Colonel William Haywood, knew of Europe's work as a band director, and called upon him to organize the 'best damn brass band in the United States Army'. Europe recruited black musicians from different parts of the United States, and in the fall of 1917 he sailed to France with his new band, nicknamed 'The Hellfighters'.[43]

For more than a year, the Hellfighters served as General John Pershing's personal band and provided entertainment for the Allied troops. Then during the winter of 1918, in response to requests from the French government, Pershing ordered the band on a six-week tour of France. Between 12 February and 29 March 1918, the Hellfighters performed in twenty-five French cities, consistently winning a passionate response from the French.[44] Much of the emotion was generated by the astonishing diversity of talent which audiences encountered when they heard Europe's band play:[45]

The Hellfighters were much more than just a brilliant brass band. Europe had included in the personnel musicians who could sing, dance, do comedy, and almost every kind of entertaining. The band could be a marching unit or break down into several dance orchestras, or into a theater band that could accompany the varied talents of the band's entertainers.

Whether the band performed as a dance, theatre, or marching unit, its playing style was syncopated and improvisatory. An

---

[42] Europe's account of the invention of the foxtrot appeared in *New York Tribune* on 22 November 1914 and is quoted by Hamm in *Music in the New World*, p. 403. W. C. Handy's version is quoted by Marshall Stearns in *Jazz Dance*, p. 98.

[43] Europe's formation of the Hellfighters is discussed in Samuel B. Charters and Leonard Kunstadt, *Jazz: A History of the New York Scene* (Garden City, NY: Doubleday, 1962), pp. 65–6.

[44] The dates of the Hellfighters' tour are provided by Charters and Kunstadt, *Jazz: A History of the New York Scene*, p. 68. There are discrepancies, however. Gunther Schuller states that Europe's band played in Aix-les-Bains and in Chamberey from 15 February to 17 March 1918, when they rejoined their regiment at the front near Giury-en-Argonne. See *Musings: The Musical Worlds of Gunther Schuller* (New York: Oxford University Press, 1986), p. 38.

[45] Charters and Kunstadt, *Jazz: A History*, p. 66.

American correspondent during the war described the forceful impact of the band's syncopated rhythms on French audiences, basing his article on an interview with Noble Sissle, drum major of the Hellfighters. The reporter quoted Sissle's description of the rhythmic excitement kindled when the drummers in the band began 'shaking in time to their syncopated raps'. Soon the audience 'began to sway', and then 'dignified French officers began to tap their feet along with the American General . . .'[46]

The most dramatic series of concerts which the Hellfighters gave in France took place in Paris. On 18 August 1918, just before the Allied Conference, Colonel Haywood asked Jim Europe to direct his band in a performance at the Théâtre des Champs-Élysées. According to Europe's own report, the Parisian audience became so 'wild' and expressed such enthusiasm during the concert that the Hellfighters ended up remaining in Paris and performing all over the city for eight weeks:[47]

Everywhere we gave a concert it was a riot, but the supreme moment came in the Tuileries Gardens when we gave a concert in conjunction with the greatest bands in the world—the British Grenadiers' Band, the band of the Garde Republicain [*sic*], and the Royal Italian Band. My band, of course, could not compare with any of these, yet the crowd, and it was such a crowd as I never saw anywhere else in the world, deserted them for us. We played to 50,000 people at least, and, had we wished it, we might be playing yet.

Jim Europe and his Hellfighters were undoubtedly the most prominent American musicians performing in France during the war. They were not alone, however, in helping to promote the French taste for American popular music. The American drummer Louis Mitchell directed a band called the Syncopating Septette in performances which took place in Paris sometime between 1915 and 1917.[48] Mitchell's band featured Crickett Smith, a talented trumpeter

---

[46] An excerpt from the article by the American correspondent is quoted by Hamm in *Music in the New World*, p. 406.

[47] This account by Europe appeared in *The Literary Digest*, 26 April 1919 and is quoted in Eileen Southern, ed., *Readings in Black American Music*, 2nd edn. (New York: W. W. Norton, 1983), p. 240. In the same article, Europe reported that the leader of the famous French band, the Garde Républicaine, was greatly impressed by the Hellfighters' performance and asked for a score of one of Europe's syncopated compositions. The French leader soon realized, however, that his band could not duplicate the rhythmic effects created by Europe's band. Europe's syncopated dance pieces may still have become part of the French band's repertoire and been performed for Parisian audiences.

[48] John Chilton states that Mitchell's band performed in Paris in the summer or fall of 1917. See *Who's Who of Jazz: Storyville to Swing Street* (New York: Time-Life Records Special Edn., 1978), p. 226. Robert Goffin is less specific and writes that between 1915 and 1917 Mitchell left London for a brief engagement in Paris with his band. See *Jazz: From the Congo*

who had worked with Jim Europe in New York before the outbreak of the war. In 1917 a newly organized band led by Mitchell appeared in Paris under the name the Jazz Kings and became a big attraction at the Casino de Paris, where they performed for the revue *Laisse-les tomber* and introduced many French art composers to jazz.[49] Along with Mitchell on drums and Crickett Smith on trumpet, the Jazz Kings offered Joe Meyers on guitar, Dan Parish on piano, Walter Kildare on bass, Frank Withers on trombone, and James Shaw on saxophone. The band performed in Paris through January of 1918 and then in the 1920s enjoyed a five-year residency at the Casino de Paris.[50] During the post-war years, Parisians were also introduced to Will Marion Cooke's Southern Syncopated Orchestra, appearing in Paris in October 1919, and, most significantly, to the great saxophone playing of Sidney Bechet who performed with the Southern Syncopated Orchestra at the Apollo in the spring of 1920.[51]

SHEET MUSIC AND SOUND RECORDINGS

In the late nineteenth and early twentieth centuries, Parisians thus had an opportunity to attend performances of American marches, cakewalks, Tin Pan Alley songs, ragtime, and syncopated dance tunes which were sung, danced to, or played by a wide variety of French and American entertainers. The French composers Erik Satie, Darius Milhaud, Francis Poulenc, and Georges Auric, who wished to become better acquainted with this American popular repertoire, could also find it published in sheet music versions and reproduced on phonograph recordings.

Sheet music of American popular tunes began circulating in France in 1891 when the International Copyright Law was passed. This law protected American and European publishers and composers of popular music against piracy and required payment to the

---

to the Metropolitan, pp. 58–9, 68. While Mitchell was directing the Syncopating Septette in Paris in 1917, he received a request from Vernon Castle to send sheet music of ragtime tunes so that he could teach the other pilots in his squadron how to dance. Mitchell obliged. In this way, ragtime music could be heard at French squadrons during the late 1910s, and sheet music was available. See Charters and Kunstadt, *Jazz: A History*, p. 68.

[49] See Goffin, *Jazz: From the Congo*, pp. 68–9.

[50] Goffin, *Jazz: From the Congo*, pp. 70, 74. According to Chilton, the performances by Louis Mitchell's Jazz Kings during the 1920s were not confined to the Casino de Paris. The band played at many other music-halls and popular institutions as well. See Chilton, *Who's Who of Jazz*, p. 226.

[51] For the information on Will Marion Cooke, see an announcement in *Variety*, 17 Oct. 1919, p. 26. Information on Bechet appears in Chilton, *Who's Who of Jazz*, p. 23.

creator or original publisher for the publication of foreign music. In order to gain access to music from abroad, domestic publishers were required to establish terms with foreign firms. Soon after the law was instituted, the French Society of Authors, Composers, and Publishers (SACEM) opened an office in New York and began to arrange copyright agreements between French and American music publishers.[52] Then on 1 July 1906, Witmark & Sons, one of the earliest and most successful of the Tin Pan Alley publishing firms, opened a small office at 92 rue St. Lazare in Paris. The firm met with such success and discovered such a large market for American sheet music that within a year and a half they were forced to move to larger quarters at 58 rue du Faubourg, Montmartre. For some time this Witmark office remained 'the only house on the Continent devoted exclusively to English and American music'.[53]

In an interview on 28 January 1908 with a reporter from the *New York Herald*, Charles Denier Warren, the manager of Witmark's Paris branch, emphasized the vital role of his office in supplying Americans in Paris (and Parisians) with the latest sheet music. Warren's description of the speed with which any popular hit in New York made its way to Witmark's Paris branch indicates that in 1908 and in the years that followed, American sheet music was readily available for Americans and Parisians alike:[54]

When an American wants a piece of music he wants it in a hurry. He has read about its success in New York, at the Broadway or Hammerstein's. It is the only thing on earth that can make him happy. Heretofore he had to send to New York for it. That meant three or four weeks. He had to guess at the price and perhaps send more than was necessary. By the time the music arrived his hunger for it had passed and the tune did not sound good to him after all. Now he gets it hot off the griddle, for he can rush in here and find any song that is making a success in New York almost as soon as he could get it there.

Parisian café orchestras also benefited from the new Witmark office; whereas in the past they had waited two or three years for sheet music of songs from American musical comedies, now they received the music as soon as the songs had become a success in

---

[52] Information on the International Copyright Law and on the New York office of SACEM provided by Russell Sanjek in correspondence. Sanjek notes that as a result of the previous very lax Copyright Law, 90% of all printed music in the 1800–25 period was of foreign origin. This high percentage persisted, with only a slight drop at the end of the century.

[53] Stated by Isidore Witmark and Isaac Goldberg in *The Story of the House of Witmark: From Ragtime to Swingtime* (New York: Lee Furman, Inc., 1939), p. 283. For further information on Copyright Laws, see p. 99.

[54] Quoted ibid., pp. 283–4.

New York.[55] Florenz Ziegfeld's 1907 production in France of Gustav Luder's *Prince of Pilsen* brought a sharp increase in the French demand for American sheet music. The operetta was performed in French, and its success prompted Witmark's Paris office to publish songs like *The Message of the Violet* and *Heidelberg Stein Song* in French and English versions. Both songs had originally been published by Witmark of New York in 1902.[56]

Since Witmark's Paris office did not open until 1906, several years separated the American and Parisian publication of songs from *Prince of Pilsen*. However, the time lag between musical successes published in New York after 1906 and their publication in Paris was very small. For this reason it is possible to obtain a fairly accurate picture of the American Tin Pan Alley tunes circulating in Paris between about 1906 and 1920.[57] The chronological list of song successes published by Witmark in New York contains music ranging from Ernest R. Ball's *Love Me and the World is Mine* (1906), an early Tin Pan Alley song, to Victor Herbert's *Italian Street Song* (1910) from the operetta *Naughty Marietta*, Louis A. Hirsch's *Hello Frisco* (1915), which was sung in the Ziegfeld Follies, and George M. Cohan's *In a Kingdom of our Own* (1919) from the operetta *The Royal Vagabond*.[58] Among these and other Witmark song publications, one may speculate that the songs most likely to have been published by the Paris firm were those which later appeared in editions by French publishers. For example, *Love Me and the World is Mine*, *Italian Street Song*, and *In a Kingdom of our Own* were all published by the Parisian firm of Éditions Feldman. Irving Berlin's overwhelmingly successful ragtime song, *Alexander's Ragtime Band*, which was published by Witmark of New York in 1911, appeared in a French edition by Salabert.[59] In addition,

---

[55] One can surmise that a café orchestra such as the one which accompanied Vernon and Irene Castle at the Café de Paris had some musical-comedy tunes in its repertoire, as well as the dance tunes introduced by the Castles.

[56] See the interview with Charles Denier Warren quoted in Witmark and Goldberg, *The Story of the House of Witmark*, p. 284.

[57] One can assume that the Paris Witmark firm remained open through 1920, though the dates are not documented. Foreign firms such as Ricordi, described below, also played a role in publishing American songs and dance music tunes, and French firms circulated editions as well.

[58] The dates which appear in parentheses represent the date of publication by Witmark. For a chronological list of Witmark's song successes, see Witmark and Goldberg, *The Story of the House of Witmark*, pp. 447–55.

[59] The information on French editions of American popular songs was obtained from Dominique Raymond, head of Documentation Générale at SACEM in Paris. The dates in the case of the latter three songs are considerably later. *Italian Street Song* was published in 1935, *In a Kingdom of our Own* in 1931, and *Alexander's Ragtime Band* in 1939. The 1939 publication of *Alexander's Ragtime Band* reflected the recent release of the American movie with the same title starring Alice Fay and Don Ameche.

certain Tin Pan Alley tunes not originally issued by Witmark in Paris were selected and published by French firms. For example, Shelton Brooks's *The Darktown Strutters' Ball* (1917), one of the most popular Tin Pan Alley tunes of its time, was published by Salabert in 1920.

The songs mentioned above represent just a sample of the dozens of American Tin Pan Alley tunes that probably circulated in Paris in both French and American editions. Witmark's Paris office was a principal source of distribution but, as the appearance of French editions of songs not originally published by Witmark indicates, it was by no means the only one. The Italian firm of Ricordi had an office in New York in the 1910s where it published American songs and dance music tunes that were then exported to Milan and distributed throughout Europe.[60] Ricordi's catalogues from the period list music published by the New York firm. Thus they offer some indication of the American popular tunes that may have reached Paris via Milan.[61] In 1914, for example, Ricordi of New York published a one-step turkey trot for piano by F. Mac Kee, a foxtrot for orchestra by Will H. Dixon, and a march two-step for military band by Jesse M. Winne. In 1915, the catalogue listed a number of dance tunes and marches by Jim Europe including *The Castle Doggy*, a foxtrot for piano; *Monkey Doodle*, a cakewalk and one-step for orchestra; and *Congratulations Waltz*. The year 1918, moreover, saw Ricordi's publication in New York of spirituals by Harry T. Burleigh and songs by Jim Europe.[62]

Dance music composed by Jim Europe and featured in the repertoire of Europe's Society Orchestra was available on record as well as in sheet music.[63] Parisians likewise had access to phonograph recordings by Sousa's band. Such records first appeared in 1900, when the International Copyright Law initiated reciprocal agreements and patents enabling American and European record manufacturers to operate. Not long after the law was passed, Victor

[60] Russell Sanjek provided me with this information on Ricordi.

[61] Ricordi did have a Paris branch entitled the Société Anonyme des Éditions Ricordi, but the catalogues of this office list almost no American popular music. The only American tune which appeared in a Ricordi Paris catalogue from June 1915 was a two-step by R. Ascione. One can assume, then, that American tunes published by Ricordi in New York arrived in Paris via Milan.

[62] See Ricordi's Milan Catalogues (which list the New York office's publications): Nos. 41, 1914; 43, 1915; 56, 1918. The songs by Jim Europe included 'Syncopated Minuet', 'Follow On'—a march song, 'Rat-a-tat drummer boy', and others.

[63] One might note that *Congratulations Waltz*, published by Ricordi of New York in 1915, was recorded by Europe's Society Orchestra on Victor in February 1914. If Ricordi did indeed export the sheet music to Paris, the possible appearance of Victor's recording in Paris may have reinforced French familiarity with the tune. For information on Victor's recordings of Jim Europe, see Brian Rust, *Jazz Records: 1897–1942*, 4th edn. rev. and enlarged, 2 vols. (New Rochelle, NY: Arlington House Publishers, 1978) i. 512–13.

and Columbia engaged the International Gramophone Company to act as their European representative. The London branch of Gramophone was renamed His Master's Voice in 1907, but in France the title of Gramophone persisted until after World War II.[64]

Once relations between American and European recording companies had been established, Victor and Columbia began sending record moulds to France and to England where Gramophone and His Master's Voice then pressed new phonograph recordings. Sometime in March or April 1903, for example, Gramophone released a recording of Sousa's band performing *At a Georgia Campmeeting*. The record was a French duplication of a disk issued by Victor in December 1902.[65] In March or April of 1914, moreover, Gramophone pressed and released a recording of Europe's Society Orchestra performing *Too Much Mustard*. The original recording had been issued by Victor in December 1913. It is noteworthy that Gramophone's release of *Too Much Mustard* coincided with the period in which Vernon and Irene Castle were popularizing this and other one-steps and turkey trots at the Café de Paris and at Deauville.

Gramophone in France did not press a large number of the early Victor recordings by Sousa and Jim Europe. However, many of these recordings were exported to France or taken abroad during the war by Americans in the Red Cross and the French Armed Services.[66] Between 1900 and 1912, Victor released over thirty recordings by Sousa's band, and in February 1914 the company issued two more recordings by Europe's Society Orchestra. The Sousa repertoire comprised coon songs, cakewalks, and early rags, and the repertoire of Jim Europe consisted of syncopated dance music. Although it is impossible to specify how many of these recordings became available in France between 1900 and 1924, it seems likely that such phonograph records provided a supplement to the abundance of live performances of American music in Paris.[67]

[64] The information on French recording companies and on French recordings was obtained from Jean-Paul Guiter of RCA, Paris.

[65] See Brian Rust, *Jazz Records*, ii. 1474.

[66] Jean-Paul Guiter and his colleague Daniel Nevers confirmed that single- and double-face recordings by Victor and Columbia were indeed exported to France. The exact quantity of records is, of course, impossible to determine. A statement on the presence of Victor recordings in France also appears in an article by M. Robert Rogers, 'Jazz Influences on French Music', *The Musical Quarterly* 21 (Jan. 1935): 56.

[67] Pathé recorded fourteen discs by the Hellfighters, but these were American recordings. Gramophone, on the other hand, issued recordings of American popular music which were not limited to Sousa and Jim Europe. The white banjo player Fred Van Eps recorded on Gramophone between 1910 and 1912; Sylvester Ossman was recorded on Victor and then pressed by Gramophone in the years 1904–6; Pietro Deiro recorded on Gramophone in 1904 and 1905. This information on Gramophone provided by Daniel Nevers and Jean-Paul Guiter.

After the war, moreover, the extreme popularity of American music was reflected in the many American band recordings made in Paris by Pathé Marconi. In March 1919 Pathé recorded an American group called the Scrap Iron Jazz Band, and between December 1921 and May 1923 they released a series of recordings by Louis Mitchell's Jazz Kings.[68]

In late nineteenth- and early twentieth-century Paris, popular amusement encompassed a rich and complex range of institutions, genres, and performers. The cabaret featured strictly Parisian entertainment, whether songs, theatrical sketches, or shadow plays; the café-concert was a centre for French popular singers; fair and circus attracted clowns, acrobats, and dancers from all over the world; music-hall featured circus performers, but placed special emphasis on its *revue à grand spectacle*; cinema offered French and American silent films. When popular entertainers from the United States visited Paris, they successfully penetrated circus, music-hall, and cinema, and performed in theatres and public squares as well. Thus Parisians with an interest in popular amusement could learn a great deal about the popular repertoire of their own country, and about American popular culture. Erik Satie, who worked in several Parisian popular institutions and frequented many others, drew his inspiration from the French and American popular entertainment which surrounded him.

[68] These early recordings can be heard on the series *Le Jazz en France*, Pathé Marconi 172 7251, vol. 1.

3

Satie and the Cabaret

THE composer and musicologist Roland-Manuel reports that he first came across the name of Erik Satie in 1910 when he learned that Debussy had orchestrated Satie's piano pieces, the *Gymnopédies*. Curious to find out more about this unknown French composer, Roland-Manuel questioned a salesman at the firm Éditions Durand. The salesman searched his memory for a few moments, and then recalled that Satie was the 'bonhomme' who had composed a cakewalk and several waltzes for the popular music-hall singer, Paulette Darty.[1] No mention was made of Satie's other compositions. Ornella Volta, editor of the French edition of Satie's writings, comments similarly upon Satie's reputation as a composer of popular song. She reports that in Arcueil-Cachan, where he lived from 1898 to 1925, Satie was considered a café-concert composer. Years later when the citizens of Arcueil discovered that he had also produced a body of art compositions, they were astonished.[2]

Both anecdotes are significant because they reveal the striking difference between Satie's role in the popular arena and the involvement of Cocteau, Milhaud, Poulenc, and Auric. Whereas the latter four were, from the start, artists who shared a spectator's enthusiasm for circus, fair, cinema, and music-hall, Satie belonged to the circle of cabaret and music-hall entertainers and led the Bohemian hand-to-mouth existence of the typical chansonnier. Whereas Cocteau, Milhaud, Poulenc, and Auric were onlookers, Satie spent the period from 1887 to 1909 both as a composer of art music (*Ogives, Sarabandes, Fils des étoiles*) and as an insider of the cabaret.[3] During these years, he produced a collection of popular pieces that reflect his mastery of a wide range of French and

---

[1] Roland-Manuel describes Satie as a composer 'unknown by my teachers and my friends'. See Roland-Manuel, 'Satie tel que je l'ai vu', *La Revue musicale* 214 (1952): 9.

[2] See Erik Satie, *Écrits*, ed. Ornella Volta (Paris: Éditions Champ Libre, 1981), p. 287.

[3] Even when Satie had ceased his cabaret work, he did not always shed the role of popular performer and composer for that of spectator. For instance, according to the singer Pierre Bertin, Satie did not join Cocteau and the band of composers, poets, and painters in their Saturday evening visits to fair, circus, cinema, and music-hall. Bertin is cited by Volta in Satie, *Écrits*, p. 303. On the other hand, Jean Wiéner mentions Satie among the crowd of artists and poets who flocked to the Bar Gaya, the predecessor of *Le Bœuf sur le toit*. See Jean Wiéner, *Allegro appassionato* (Paris: Pierre Belfond, 1978), p. 44.

American styles, from cabaret song to *valse chantée*, march, and cakewalk. Even after ceasing his work for popular institutions, Satie continued to derive inspiration from Parisian entertainment; the piano pieces with humorous annotations and the essays on music which first appeared in the 1910s contain a blend of fantasy, wit, and parody that clearly stems from the Parisian cabaret. In the late 1910s and early 1920s, moreover, Satie incorporated cakewalk, waltz, and march into the fabric of such large-scale compositions as *La Belle Excentrique, Parade, Mercure*, and *Relâche*. Thus the sounds of Parisian entertainment reverberate throughout Satie's career. This chapter traces Satie's early participation in cabaret, café-concert, and music-hall activities, as well as the absorption of cabaret humour into his writings.

ACCOMPANIST AND COMPOSER FOR PARISIAN VENUES

According to his friend, the poet Contamine de Latour, who accompanied Satie on his early visits to cabarets, it was the Chat Noir which 'revealed to him his vocation and transformed him completely'.[4] Satie began frequenting the Chat Noir in 1887. The cabaret's atmosphere of caustic wit and mockery appealed to him immediately, and he joined the *poètes-chansonniers* and painters in their defiance of convention and their flippant disregard for the future. The painter Henri Rivière, the poets Jules Jouy and Maurice Donnay, and the poet and humorist Alphonse Allais were just a few of the cabaret performers whose company Satie enjoyed.[5] He and Allais shared a special bond. They were both born in the town of Honfleur, where they attended the same college, and their families had been acquainted for some time.

Satie and his *chansonnier* friends boldly challenged bourgeois conventions by inventing absurdist songs and plays and writing humorous satires in the *Chat Noir* journal. Their provocative, anti-art spirit prefigured in a remarkable way the Dada spirit which invaded post-war Paris in 1919 and held sway until 1921. Indeed Dada's origins in a Zurich cabaret called the Cabaret Voltaire (1916) reinforce the movement's indebtedness to a cabaret aesthetic—French as well as German. During his lifetime, Satie drew a mixed

[4] See P. Contamine de Latour, 'Erik Satie intime: souvenirs de jeunesse', *Comoedia*, 3 Aug. 1925, p. 4. Translations here and throughout the chapter are my own unless otherwise indicated.

[5] For the atmosphere at the Chat Noir, see 'Souvenirs de jeunesse', 3 Aug. 1925. Contamine de Latour also describes the friendship between Satie and Alphonse Allais. Pierre-Daniel Templier discusses Satie's acquaintances at the Chat Noir in *Erik Satie* (Paris: Rieder, 1932), p. 14.

reception from the Dadaists; *Parade*, for example, received strong criticism from Dadaists attending the May 1917 première.[6] Yet the absurd, satiric elements in Satie's writings and compositions intrigued such artists as Francis Picabia and Man Ray. The stuffed monkey who dances quadrilles and the baron who talks on the phone to a horse in *Le Piège de Méduse* (1913); the 'Perpetual Tango' from *Sports et divertissements* (1914) in which the performer is instructed to play the same dance over and over again; the typewriter and Morse apparatus in *Parade*—all of these elements anticipated the Dadaist debunking of artistic conventions. In 1951 the American painter Robert Motherwell gave serious recognition to Satie as a precursor of Dada by including two passages from the composer's 'Memoirs of an Amnesiac' (1913) in the 'Pre-dada' section of his pioneering and comprehensive Dada anthology. Significantly, the eccentric humour and the parodies of academicism which characterize both these literary sketches and Satie's music had their source in the lively discussions and collaborations with poets and painters at the Chat Noir.

In about February of 1888, Satie became assistant to the principal pianist of the Chat Noir, Albert Tinchant. Satie's activities involved accompanying singers, writing occasional song arrangements, and playing piano for the shadow plays.[7] He continued this work until 1891 when he left the Chat Noir and became the pianist at another cabaret, the Auberge du Clou. According to Nigel Wilkins, the engagement at the Auberge du Clou was brief; by 1892 Satie had taken a post at the Café de la Nouvelle Athènes.[8] The length of Satie's stay at this café and the exact nature of his activities are not known. After the Nouvelle Athènes, Satie no longer held positions as the regular pianist at an individual popular establishment.

[6] For information on Dada's origins and its appearance in Paris, see Georges Ribemont-Dessaignes, 'History of Dada', in Robert Motherwell, ed., *The Dada Painters and Poets: An Anthology*, 2nd edn. (Cambridge, Mass.: The Belknap Press of Harvard University Press, 1989), p. 108. Decades later, Breton apparently told Motherwell that surrealism was against music. Such attitudes may explain Satie's comment, 'Everyone will tell you that I am not a musician.' See Motherwell, ed., *The Dada Painters*, p. xxiii.

[7] The date of 1888 is suggested by Roland Belicha in 'Chronologie Satiste ou photocopie d'un original', *La Revue musicale* 312 (1978): 14. The *Chat Noir* weekly journal never mentioned Satie's cabaret work. Performances involving Albert Tinchant, however, were often announced, and one can assume that Satie's activities duplicated Tinchant's. For this information, see Steven Moore Whiting, 'Erik Satie and Parisian Musical Entertainment 1888–1909', Master's Thesis (Musicology), University of Illinois, Urbana, 1983, pp. 55–6.

[8] These dates according to a chronology in *The Writings of Erik Satie*, ed. Nigel Wilkins (London: Eulenburg Books, 1980), pp. 20–1. Templier provides the same date for Satie's departure from the Chat Noir in *Erik Satie*, p. 19. Wilkins's view on Satie's post at the Nouvelle Athènes is corroborated by Roland-Manuel who reports that Maurice Ravel's father first introduced his son to Erik Satie at the Nouvelle Athènes, where he knew that Satie could always be found.

Rather, he worked in several different popular milieux and shifted the focus of his activities from performance to composition. A position which exemplifies this transition is Satie's dual role as pianist and composer of waltzes and other music-hall songs at a Montmartre cabaret called La Lune Rousse or The Harvest Moon.[9] In 1899 Satie began collaborating with the *poète-chansonnier* Vincent Hyspa, who made nightly tours of the Montmartre cabarets. Satie served first as Hyspa's pianist and by 1900 as his composer.

The music notebooks of Satie which are preserved in the Houghton Library of Harvard University contain twenty-eight songs with newly composed music by Satie and lyrics by Vincent Hyspa.[10] Of these, only two—'Un Diner à l'Élysée' and 'Chez le Docteur'— have been published. The melody of the first appeared in a 1903 anthology of Hyspa's songs entitled *Chansons d'humour*, and the melody and accompaniment of the second were published in 1976 by Éditions Salabert. 'Un Diner à l'Élysée' merits attention because its application of irregular phrasing to a simple tune represents a stylistic device that Satie applied in many of his art compositions— both those from the same period as 'Un Diner' and those composed later. For instance, asymmetrical phrasing shapes the café-concert tune in 'Prolongation du même' from *Trois Morceaux en forme de poire* (1903), the chorale in *Sports et divertissements* (1914) and the 'Steamship Ragtime' in *Parade* (1917).

The topic selected by Hyspa for his lyrics was a banquet in which the President of the French Republic invites members of the Société des Artistes Français and the Société Nationale des Beaux-Arts. The text pokes fun at the disparity between the President's pretensions to hospitality and decorum, and the dearth of liquor and meagre conversation. The song's refrain continues this satire on French nationalism and on the French government's lack of munificence towards French artists, by juxtaposing a musical quotation from 'La Marseillaise' against a line of spoken text in which the singer sarcastically praises the French national anthem for being 'vraiment français'.

---

[9] Virgil Thomson mentions Satie's work at The Harvest Moon in a 1941 *New York Herald Tribune* article, 'French Music Here', reprinted in *A Virgil Thomson Reader* (Boston, Mass.: Houghton Miflin Co., 1981), pp. 207–8. No dates are provided.

[10] This material and the information on Satie's shift from accompanist to composer are based on Steven Whiting's study of unpublished manuscripts of popular songs arranged and/ or composed by Satie for Hyspa between around 1899 and 1903. See Whiting, 'Erik Satie and Parisian Musical Entertainment', p. 159.

I

Le Président, d'une façon fort civile,
Avait invité nos grands peintres français
À venir goûter de sa cuisine à l'huile.
    On raconte que ce fut vraiment parfait.
        Après la soupe, radis et caviar
        Pour faire plaisir au Czar . . .
Ça sentait bon—et le moment était suprême—
    Et la musique du soixante-quatorzième
        De ligne jouait,
        (Ne vous déplaise),
        La Marseillaise,
    Hymne vraiment français
        (Ou française).

II

La conversation avait été tres maigre
jusque-là, quand l'épouse du Président,
    qui avait à sa gauche ce sale Leygues
Et à sa droite Monsieur Jean-Paul Laurens,
Dit tout à coup au Ministre des Beaux-Arts:
    'En voulez-vous, du z-homard? . . .'
        Ça sentait bon . . .

III

Mais subitement les liquides manquèrent
    (On en était au gigot aux haricots).
Le Président dit à son fils: 'Ventre à terre!
Cours chez notre bistro de la rue Duphot
Me chercher douze bons vieux litres de choix
. . .
    Dis-lui bien que c'est pour moi . . .'
        Ça sentait bon . . .

IV

Après le café—ce grand noircisseur d'âmes,—
    Ces messieurs assurèrent sérieusement
Notre Président Auguste (pour les dames)
    Et sa dame, de leur profond dévouement,
Puis ils se retirèrent tranquillement
    Et tout en borborygmant . . .
        Ça sentait bon . . .

(I) The President, in an extremely polite fashion, | had invited our great French painters | to come and sample his gourmet cuisine. | People say that it was truly perfect. | After the soup, one had radish and caviar | in order to please the Czar . . . | It tasted good—and the moment was ideal— | and the music of the 74th regiment played, | (don't be offended) | The Marseillaise, | Hymn truly *français* | (or *française*). (II) The conversation was very meagre | until the President's wife, | who had this dirty Mr. Leygues on her left | and on her right Mr. Jean-Paul Laurens, | said all at once to the Minister of Beaux-Arts: | 'Would you like some lobster? . . .' | It tasted good . . . | (III) But suddenly there was nothing more to drink | (one was on the course with mutton and beans). | The President said to his son: 'Quick! Quick! | Run to our local pub on Rue Duphot | and find me twelve good ripe choice litres . . . | Tell him that it's for me . . .'' | It tasted good . . . (IV) After the coffee—that great substance that blackens our souls,— | these messieurs earnestly assured | our President Augustus (Augustus for the ladies) | and his wife, of their profound devotion, | and all the while making rumbling noises in their stomachs . . . | It tasted good . . . [Translation by Nancy Perloff]

The music which Satie composed for the verses of 'Un Diner à l'Élysée', best studied from the Houghton manuscript which contains the accompaniment as well as the melody (see Ex. 3.1),

Ex. 3.1. "Un Dîner à l'Elysée", transcribed from the manuscript at the Houghton Library, Harvard University

bears certain resemblances to march music. The piece is written in a clear duple metre and features an accompaniment in which octaves on strong beats in the left hand alternate with offbeat chords in the right. The melody recalls march tunes in its major mode and in its use of ascending and descending leaps of perfect fourth and fifth (see bars 20–3) which suggest a brass fanfare.

'Un Diner à l'Élysée' departs both from march and from popular song styles, however, in its asymmetrical phrasing. Each strophe contains two four-bar phrases followed by an eight-bar phrase, a four-bar phrase, and a two-bar closing idea. The only exact melodic repetition occurs in the last four-bar phrase (bars 21–4), which subdivides into two identical units. Otherwise the melody presents ongoing new material, without even the use of sequence.

Yet despite the asymmetry, Satie's song is tuneful because melodic movement is confined to steps, thirds, and an occasional fourth or fifth, and successive pitches are frequently repeated (see bars 13–16). The rhythm, moreover, scarcely strays from short motifs comprised of repeated quarter notes and eighth notes. In this way, Satie creates a melody of utmost simplicity which permits a clear, syllabic rendering of the text. He occasionally calls attention to the satiric aims of Hyspa's lyrics by displacing the accent in a word like 'français' from the second syllable to the first (bar 12). The resultant sound, which is abrupt and comical, reinforces Hyspa's mockery of the French President's pretensions to patriotic sentiment and to support for his country's artists.

In addition to Satie's cabaret work with Vincent Hyspa, he composed music-hall songs for the singer Paulette Darty. Two waltzes, 'Je te veux' and 'Tendrement', date from about 1902, and a song in the style of a syncopated cakewalk, 'La Diva de l'Empire', was composed two years later.[11] The sung waltz or *valse chantée* was

[11] 'Tendrement' and 'Je te veux' were deposited at the Société des Auteurs in 1902, on 29 March and 26 November respectively. See Belicha, 'Chronologie Satiste', p. 20.

Paulette Darty's special invention; Chantal Brunschwig, an historian of French popular song, describes how Darty's performances of waltzes at the Scala and the Ambassadeurs in the 1890s and 1900s won her the name of 'reine des *valses lentes*' ('queen of slow waltzes'). Contemporary reviews in the entertainment column of *Le Figaro* corroborate Brunschwig's statement, praising Darty's 'charming and supple voice' and calling her waltzes 'tender, lilting [pieces] which assume the demeanour of a secret or a timid confession'.[12] Most of Darty's *valses chantées* were composed by the Viennese *émigré* Rodolphe Berger, with lyrics by Maurice de Féraudy. Ex. 3.2 contains the opening of a *valse chantée* from Berger's operetta, *La Femme de César*. Berger wrote in a slow, languid, sentimental style which distinguished his waltzes from their Viennese models.[13]

When Satie turned to the *valse chantée*, his music assumed a

Ex. 3.2. Opening of Rodolphe Berger, *valse chantée*, from *La Femme de César*

---

[12] See Chantal Brunschwig, *100 ans de chanson française* (Paris: Aux Éditions du Seuil, 1972), p. 112. For contemporary reviews, see *Le Figaro*, 15 May 1900; 31 May 1900.
[13] Mosco Carner, *The Waltz* (London: Max Parrish, 1958), p. 58.

romantic tone similar to Berger's and strikingly different from the light, humorous style of his cabaret songs. This contrast is especially apparent in 'Tendrement', in which the lyrics are once again by Vincent Hyspa but bear no trace of the satire and irony that pervaded 'Un Diner à l'Élysée'. Rather, they are excessively sentimental, as the following words from the refrain demonstrate, and serve principally to reinforce the emotional mood of the music:

> D'un amour tendre et pur
> Afin qu'il vous souvienne,
> Voici mon cœur, mon cœur tremblant,
> Mon pauvre cœur d'enfant
> Et voici, pâle fleur
> que vous fites éclore,
> Mon âme qui se meurt de vous
> Et de vos yeux si doux.

> A love tender and pure
> So that you may remember it,
> Here is my heart, my trembling heart,
> My poor heart of a child
> And here, pale flower,
> So that you blossom,
> Here is my soul which is dying for you
> And for your eyes so sweet.

Satie's melody acquires its wistful, romantic quality through occasional chromatic inflection, and through the use of ties within and across the bar (see Ex. 3.3*a–b*) so that rubato is incorporated into the rhythmic fabric. Both devices were commonly found in waltz music. Satie duplicates the form of Berger's *valses chantées*, moreover, by creating an alternation between a thirty-two-bar refrain in the tonic and a thirty-two-bar verse in the dominant. The frequent use of ties, the division of verse and refrain into two sixteen-bar strains, and the design of refrain followed by verse were traits that also appeared in Satie's second waltz, 'Je te veux'.

Paulette Darty did not confine her talents to the performance of *valses chantées*. In 1904 she sang Satie's cakewalk song, 'La Diva de l'Empire', in a music-hall revue by Dominique Bonnaud and Numa Blès entitled *Devidons la bobine* (Let's Unwind the Bobbin).[14] The composition of this song for Darty demonstrates both the singer's and the composer's impressive versatility and mastery of a wide range of popular styles.

---

[14] James Harding discusses Satie's composition of 'La Diva' for a music-hall revue in *Erik Satie* (New York: Praeger, 1975), p. 78.

Ex. 3.3. The incorporation of rubato in "Tendrement"

(a)

vien - ne, Voi - ci - mon coeur, mon coeur _ trem-blant,

(b)

Mon â - me qui se meurt _ de vous Et de _ vos yeux si

au 2ᵉ couplet
aller au signe

p reten.

    Satie's choice of a syncopated cakewalk expresses the Parisian
enthusiasm for the abundance of cakewalk performances and
contests taking place in Paris between 1900 and 1904.[15] His
application of a march tempo suggests his likely understanding of
the close resemblance between marches and early syncopated
cakewalks; the label 'march' appeared frequently as a subtitle on
sheet music of cakewalks in the 1890s.[16] It was Satie's use of 'untied
syncopation' in the melody, however, that identified 'La Diva'
immediately as a cakewalk:

    This rhythmic pattern, a pervasive feature of such cakewalks as *At
a Georgia Campmeeting* and *Smoky Mokes* which were performed

---

[15] Debussy also responded with his 'Golliwog's Cakewalk' from *Children's Corner* (1906–8).
[16] Edward Berlin discusses the relationship between cakewalk and march in *Ragtime: A Musical and Cultural History* (Berkeley: University of California Press, 1980), pp. 104–5.

by Sousa's band, imposes syncopation in separate halves of the bar rather than across the middle of the bar. The offbeat accentuations are strongly felt because left-hand articulations appear on strong and weak beats, and the syncopated pitches of the melody fall between them.[17] Ex. 3.4*a–b* illustrates two instances of untied syncopation in 'La Diva'. In Ex. 3.4*a*, Satie applies the pattern to the melody without providing punctuations on strong beats in the left hand to enhance the syncopated effect. In Ex. 3.4*b*, by contrast, steady eighth notes in the left hand reinforce the syncopation.

Ex. 3.4. Untied syncopation in "La Diva de l'Empire"

---

[17] Edward Berlin introduces the term 'untied syncopation' and describes the pattern in *Ragtime*, p. 83. Cakewalks can be distinguished from tangos by the cakewalk's use of untied syncopation in the melody and the tango's use of this pattern in the accompaniment. Although Satie introduced untied syncopation in the melody of his 'Perpetuel Tango' from *Sports et divertissements*, he also used the accompaniment pattern of dotted eighth followed by a sixteenth and two eighths (a typical tango accompaniment), as opposed to the steady left-hand eighth notes of the cakewalk.

In several bars of his introduction to 'La Diva', Satie omits the left-hand accompaniment and doubles the melody at the octave (see Ex. 3.5*a*). This choice of unharmonized octaves, like the untied syncopation, reflects Satie's familiarity with American cakewalks and early rags. Unharmonized octaves often occupied at least two bars of the typical cakewalk introduction, as shown in Ex. 3.5*b*. 'La Diva' also reveals Satie's awareness of the tendency in American cakewalk introductions to use both octave doubling and linear chromaticism (see Ex. 3.5*a*, bar 3). The combination of octaves and linear chromaticism can be traced to the introduction of Abe Holzmann's *Hunky-Dory*, which was also performed by Sousa in Paris (see Ex. 3.5*c*).

Ex. 3.5. Unharmonized octaves and linear chromaticism in "La Diva de l'Empire" and two American cakewalks

(a) "La Diva de l'Empire" Introduction

(b) At a Georgia Campmeeting Introduction

(c) Hunky Dory Introduction

In the same year that Satie wrote 'La Diva de l'Empire', his fascination with American cakewalks and rags inspired him to compose an instrumental cakewalk which he called *Le Piccadilly*. Satie scored the piece for solo piano and, alternatively, for a small string orchestra and subtitled it 'marche', just as he had assigned a march tempo to the music of 'La Diva'. *Le Piccadilly* did not appear in published form, however, until 1975 when Éditions Salabert issued the piano version. Three years later, Salabert published Satie's orchestral arrangement and advertised it in their catalogue as an *orchestration de l'epoque* ('period orchestration') for an *orchestre de brasserie* or a 'pub band'.[18] The designation *orchestre de brasserie* provides some clue about the original function of *Le Piccadilly*. The dance may have been played as background music at the café-concert or during interludes between spectacles at the music-hall. Certainly the title evokes circus and, by extension, music-hall entertainment.

Stylistic traits of ragtime that appear in 'La Diva de l'Empire' assume greater prominence in Satie's instrumental cakewalk. Untied syncopation, which Satie applies less frequently than dotted rhythms in 'La Diva', is heard in twenty-two of the forty-four bars of *Le Piccadilly*. Monophonic octave passages, used sparingly in the introduction of 'La Diva', occupy four of the eight bars of Satie's introduction to *Le Piccadilly* and are combined once again with linear chromaticism (see Ex. 3.6). Satie also observes the practice typically found in syncopated cakewalks like *At a Georgia Campmeeting* of following the opening strains with a 'trio' which modulates to the subdominant.

Ex. 3.6. Unharmonized octaves and linear chromaticism in *Le Piccadilly*

---

[18] The present author was able to peruse a copy of the orchestral version of *Le Piccadilly*, which G. Schirmer obtained from Salabert. Satie scored the piece for violins (divisi), violas, cellos, double bass. Although Satie's waltzes and his cakewalk for Darty were also arranged for *orchestre de brasserie*, and may (like *Le Piccadilly*) have functioned as music-hall background music, they were written primarily for performance by Darty. In the case of *Le Piccadilly*, background music for café-concert or music-hall was probably the primary function, since we know of no other reason for the composition of this piece.

The formal design of *Le Piccadilly* does not coincide with the characteristic scheme of American cakewalks, however. Whereas the majority of cakewalks and early rags contain three or four different strains, each comprised of sixteen bars and featuring a new theme, Satie limits the number of strains to two. Thus the piece follows the design: Introduction A (8 bars), Strain A, Introduction B (4 bars), Strain B (Trio), Reprise of Introduction A and Strain A. The procedure of immediate repeats of each strain, nearly always observed in early rags, also does not apply. Indeed the use of only two strains and the omission of immediate repeats aligns the form of *Le Piccadilly* more closely with the song form of 'Tendrement' and 'La Diva de l'Empire', both of which alternate between refrain and verse. Yet the design of *Le Piccadilly* remains distinct from either cakewalk or song, since Strain B (analogous to the verse) is heard just once.

Most of Satie's music for cabaret, café-concert, and music-hall was composed between 1900 and 1904 in collaboration with Vincent Hyspa and Paulette Darty. During these years, Satie also completed his now celebrated art composition *Trois Morceaux en forme de poire*, portions of which originated as café-concert pieces.[19] Several projects after 1904 attest to Satie's continued work in popular composition. In 1905, for example, one year after he had written 'La Diva de l'empire' and *Le Piccadilly*, Satie began collaborating with the French popular lyricist Maurice de Féraudy on a comic operetta, *Pousse l'amour*, which was probably intended for café-concert or music-hall. Satie composed several numbers and then became involved in disagreements which led to his withdrawal from the project. The completed operetta was performed in Monte-Carlo in 1913 under the title *Coco cheri!*[20]

Collaborations with Paulette Darty and Vincent Hyspa in 1909 proved more successful. On 17 October 1909 the local newspaper of Arcueil-Cachan, *L'Avenir d'Arcueil-Cachan*, announced a 'matinée artistique' to take place on 24 October, featuring Darty and Hyspa in a performance of works by Satie.[21] The concert was one of many community activities which Satie organized as a means of raising money for local charities. Through such contributions, he became extremely well-known and well-liked in the suburb of Arcueil, both as a sponsor of community concerts and youth club outings and as a

[19] For information on the popular sources, see William Patrick Gowers, 'Erik Satie: His Studies, Notebooks, and Critics', Ph.D. (Musicology), Cambridge University, June 1965, pp. 118–46.

[20] Information on *Pousse l'amour* provided in Templier, *Erik Satie*, p. 27.

[21] This announcement is reprinted in *The Writings of Erik Satie*, p. 158. Templier states that Hyspa and Darty performed music by Satie. See *Erik Satie*, p. 29.

café-concert composer. *L'Avenir d'Arcueil-Cachan*, for which Satie often wrote, reported regularly on his musical activities. On 21 November 1909, the newspaper offered the following review of a concert at the Scala in which Paulette Darty premièred a new work by Satie and lyricist Jules Dépaquit:[22]

Mlle Paulette Darty, at the Scala, has recently given the first performance of a delightful fantasy by Dépaquit called 'La Chemise', which our friend Satie has embroidered with lively and graceful tunes which will soon be on everyone's lips.

Every evening Performer, Author and Composer receive deserved acclamation.

Although it is not certain whether 'La Chemise' was Satie's last music-hall composition, no further documentation of popular pieces exists.[23] The years 1900–9 most likely define the period, then, in which Satie's work as a composer for cabaret, café-concert, and music-hall spawned a varied repertoire written in the style of French and American popular songs and dances, and in collaboration with prominent singers and lyricists in Paris.

CABARET HUMOUR IN SATIE'S WRITINGS

The bulk of Satie's writings coincide not with his years as cabaret accompanist or with his period of popular composition, but with his emergence as an art composer in the 1910s. In 1911 Maurice Ravel gave the first performance of Satie's *Sarabande* No. 2, his *Gymnopédie* No. 3, and his 'Prelude' from *Le Fils des étoiles*, at the Société Indépendante de Musique. In 1912 Satie began publishing literary sketches and essays which offered a witty, ironic snapshot of his musical career and of musical life in Paris. In addition, he composed the first in a series of miniature piano pieces adorned with eccentric texts. The simultaneous appearance of the writings and the piano pieces with incidental texts, only one year after Ravel and a group of young French musicians had acknowledged Satie's importance as an art composer, suggests that public recognition gave Satie the courage and the audacity to embark on a career as a writer and musical humorist.[24]

---

[22] The review is reprinted in *The Writings of Erik Satie*, p. 160. Jules Dépaquit was a Montmartre friend who also wrote the scenario for Satie's musical pantomime *Jack in the Box* (1899).

[23] The manuscript of 'La Chemise' appears in the Houghton collection.

[24] The only essays of Satie that predate the 1910s are his mock-religious pamphlets, written between 1893 and 1895 on behalf of his imaginary church, l'Église Métropolitaine d'Art de Jésus Conducteur, and his short blurb entitled 'Montmartre Musicians', which appeared in the *Guide de l'étranger à Montmartre* in 1900. Humorous annotations and titles in his music can be traced to the *Gnossiennes* (1890) and the *Pièces froides* (1897), but became particularly prevalent in the 1910s.

Satie's wit bears a striking resemblance to the satiric tone of the *Chat Noir* journal, and particularly to the humour of his friend and colleague at the Chat Noir, Alphonse Allais. Both men delighted in absurd self-portraits which lavished excessive praise or mockery on one's personal situation and achievements, in parodies of pedantry and academicism, and in the adoption of poses as a means of caricaturing certain human types. Since Allais's writings predate Satie's by about twenty years, it seems likely that they offered an important model for the composer. The satiric debunking of academicism and the delight in absurdity as a means of challenging the meaning of art offer uncanny harbingers of Dadaism which burgeoned in Paris in 1919.

Alphonse Allais began writing anecdotes and commentaries for various Parisian literary revues while completing his studies in pharmacology. In 1880, after he received his degree, he immersed himself in the Parisian literary world. Allais's principal activity centred around the *Chat Noir* journal. He contributed stories, and between 1886 and 1892 he served as editor-in-chief. Allais's writings for the journal range from parodies of Parisian newspaper reviews, which he signed with the name of a music critic for *Le Temps*, Francisque Sarcey, to personal narratives, and vignettes about Parisian life. The following announcement, which Allais wrote for the *Chat Noir* on 19 May 1883, is an example of one of his personal narratives:[25]

Alphonse Allais, English apothecary, has just discovered a new species of shrub destined to render great services to the sex of Joffrin.

The eminent naturalist has named his interesting discovery Parvula Nux Baudruchifera.

The renowned pharmacist, hardened philanthropist, has immediately resolved to benefit his contemporaries with the result of his learned research by organizing an insurance company to protect against secondary and tertiary accidents, under the presidency of Dr. Ricord.

People must call him up!

The reference to himself in the third person, the use of a deadpan style and a pompous, flowery language that parody newspaper writing, the accumulation of descriptive phrases glorifying his achievements ('eminent naturalist', 'renowned pharmacist'), and the element of absurdity (the organization of an insurance company) were all taken up with similar comical effect by Satie in his 1913 self-portrait for the publisher E. Demets. Instead of 'learned research'

---

[25] For biographical information on Allais, see the preface to Allais's collected writings: Alphonse Allais, *Œuvres posthumes*, ed. François Caradec and Pascal Pia, 8 vols. (Paris: La Table Ronde, 1966) i. 9. Allais's personal narrative, which first appeared in the *Chat Noir* on 19 May 1883 under the title 'Reclame à payer', is reprinted in *Œuvres posthumes* i. 164.

of a 'renowned pharmacist', we read about 'lofty genres' of a 'precious composer':[26]

M. Erik **Satie** was born in Honfleur (Calvados) on 17 May 1866. He is said to be the strangest musician of our time. He classifies himself among 'whimsical' writers who are, he says, 'nice, decent people'. . . .

We should not forget that the master is considered, by a large number of 'young' musicians, as the precursor and apostle of the musical revolution now in progress . . .

Having discussed the more lofty genres, the precious composer now explains his humorous works. This is what he says about his humour:

—'My sense of humour recalls that of Cromwell. I also owe much to Christopher Columbus: for American wit has sometimes tapped me on the shoulder and I have been pleased to feel its ironical and icy bite. . . .

The beautiful and limpid *Aperçus désagréables* [Nasty Glimpses] (piano duet: Pastorale, Chorale, and Fugue) are most elevated in style and show how the subtle composer is able to say:

—'Before I compose a piece, I walk round it several times, accompanied by myself.'

Satie's frequent satires of pedantry can also be traced to the satires of Allais. On 11 October 1884, Allais contributed an article to the *Chat Noir* entitled 'Chronique cholériforme', in which he imitated the tone of an excited journalist announcing a new breakthrough in medicine.[27] Humour in Allais's chronicle stemmed from the disparity between hyperbolical praise of these medical findings and their inconsequential nature. Thus the reader learned that the possibility of cholera being transmitted to rabbits was 'one of the most reassuring things of this century', that it possessed an importance which 'will escape no one'. Allais also praised the 'learned doctors' for restoring the reputation of 'congenial' bacteria which had, according to Allais, been unjustly vilified. This humorous anthropomorphism, intended by Allais as a satire on the dryness of scientific research, can be linked to a mock-scientific discussion of sound measurement which likewise parodied pedantry and provided the central issue of Satie's self-portrait, 'What I Am', written in 1913 as part of 'Memoirs of an Amnesiac' for the *Revue de la Société Internationale de Musique*. Here Satie coined his own scientific terminology to describe the measurement of musical pitches. Just as Allais had personified the germ, Satie endowed the pitch with physical dimensions and spoke of it as a visible object:[28]

[26] This self-portrait translated and reprinted in *The Writings of Erik Satie*, p. 79.
[27] *Chat Noir*, 11 Oct. 1884, reprinted in Allais, *Œuvres posthumes* i. 172–3.
[28] For a translation of 'What I Am', see *The Writings of Erik Satie*, p. 58.

The first time I used a phonoscope, I examined a B flat of medium size. I can assure you that I have never seen anything so revolting. I called in my man to show it to him.

On my phono-scales a common or garden F sharp registered 93 kilos. It came out of a fat tenor whom I also weighed.

Do you know how to clean sounds? It's a filthy business. Stretching them out is cleaner; indexing them is a meticulous task and needs good eyesight. Here, we are in the realm of phonotechnique.

For Satie, neologisms were a means of satirizing the arcane vocabularies and languages of academicians. In 'What I Am', his invention of terms like 'phonoscope' and 'phonometer' and his arbitrary substitution of one such term for another, mocked the precision of scientific nomenclature. In his set of piano pieces from 1912, *Véritables Préludes flasques*, Satie replaced traditional performance directions and expressive markings with Latin neologisms. Thus fictitious words such as 'Caeremoniosus', 'Paedagogus', and 'Corpulentus' appeared above the staff, in mockery of musicians who, in Satie's opinion, took themselves and their music too seriously. Similar performance directions written with humorous intent, and nearly impossible to realize musically—for example, 'à la carcasse' (in the manner of a skeleton), 'physionomique' (physiognomical), 'gluant' (sticky), 'buvez' (drink—[imperative])—appeared later in Satie's two-piano version of *Parade*.

Allais's pose of the moralist lamenting the depravity of the world also resurfaced in the writings of Satie. In January of 1889, Allais wrote an article for the *Chat Noir* in which he expressed an abhorrence for café life because it took one so far away from 'study and prayer'. 'How I prefer the powerful harmonies of [pipe] organs!', he wrote sarcastically.[29] Several paragraphs later the reader found Allais ensconced at a local brasserie where, he explained, he had been 'forced' by the pressures of modern life. In 'Painful Examples' (1922), Satie created a similar kind of irony. After heaping vehement attacks on taverns and calling them 'sad aperitival exhibitions', 'public bacchanalia', 'horrendous intemperacies' which 'constitute an offence against Morality', he remarked in the same breath, 'Obviously, I do occasionally find myself in a Pub; but I hide in the background . . .' The closeness of Satie's sarcastic, satiric point of view to Allais's is made particularly evident by Satie's introduction of his friend's name. After confessing how ashamed he should feel to be seen at a café, Satie writes, 'As Alphonse Allais used to tell me: "You could have your chances of marriage spoiled by something like that".'[30]

[29] *Chat Noir* 63 (26 Jan. 1889): 1265.
[30] For translation of 'Painful Examples', see *The Writings of Erik Satie*, pp. 121–2.

If the similarities between the writings of Satie and Allais are especially noteworthy, Satie's humour can also be compared to that of many other cabaret entertainers and writers. A perusal of anecdotes and notices in the *Chat Noir* reveals that absurd poses, hyperbole, and satires of academic seriousness were essential ingredients of cabaret humour. The essay on Montmartre which opened the first issue (14 January 1882) of the *Chat Noir*, for example, pronounced Montmartre the 'cradle of civilization' and described how the first dry land which Noah discovered after the flood was the soil of Montmartre. After making further extravagant claims about Montmartre's position at the 'centre of the world', the author of this revisionist history, Jacques Léhardy, emphasized the importance of both 'ethnographic' and 'philologic' considerations in a study of Montmartre. Léhardy's hyperbole, his pseudo-scientific references, and his careful use of a straightforward, deadpan style were humorous devices indulged in with equal glee by Allais and Satie. Indeed it is interesting to compare Léhardy's claims about Montmartre as the centre of the world, to Satie's maxim that 'the centre of Paris is France—with her colonies, naturally'.[31] Both statements derive their humour from the relegation of France to a minor position *vis-à-vis* centres like Montmartre and Paris.

The particular humour of Satie's writings reflects the importance of the cabaret as a source of influence on his aesthetic. Satie's view of Parisian popular entertainment must be studied primarily in terms of this humour, for he scarcely ever discussed his view of the popular world. A few statements appearing in letters to his brother Conrad in 1899 and 1900 attacked the popular compositions he was then writing as 'de rudes saloperies', or vulgar pieces, which he churned out only in order to make a living. In comments made after the success of *Parade*, by contrast, Satie defended popular entertainment. For instance, in one of his many aphorisms he cautioned his fellow-composers, 'Do not forget what we owe to the Music-Hall, to the Circus. It is from there that stem the newest creations, tendencies, and curiosities. The Music-Hall, the circus possess the "esprit nouveau".'[32] In 1924, moreover, in the programme of the ballet *Relâche*, Satie included a statement in which he upheld his use of popular themes and defied any 'moralist' to criticize the everyday mood of his music: 'The music of *Relâche*? I depict in it people who are gallivanting. In order to do that, I have made use of popular themes. These themes are naturally evocative . . . The "timid", and

---

[31] Satie, *Écrits*, p. 165.
[32] For Satie's statement on his 'vulgar pieces' and his aphorism on the music-hall, see *Écrits*, pp. 256, 164.

other "moralists" reproach me for the use of these themes. I don't have to concern myself with the opinion of such people.'[33]

Yet given the scarcity of essays in which Satie openly referred to his work for cabaret and café-concert and the ironic intentions of his writings, it is impossible to know how seriously he wished these attacks and defences to be taken. Satire may have played as important a role here as in the humorous essays and anecdotes discussed above.

Cocteau, Milhaud, Poulenc, and Auric were enthusiastic spectators of Parisian amusement, rather than participants. Their writings, unlike Satie's, abound in references to favourite performers and music-hall acts and to values of nostalgia, simplicity, diversity, and parody which they discovered in popular entertainment. A study of these writings reveals important similarities in point of view, as well as a desire to embrace some of the same principles of the Parisian popular world—parody, impiety—which attracted Satie.

[33] Grete Wehmeyer includes a reprint of the programme of Satie's *Relâche*, where the statement on popular themes appeared. Wehmeyer, *Erik Satie* (Regensburg: Gustav Bosse Verlag, 1974), p. 276.

4

The Popular World of Cocteau, Milhaud, Poulenc, and Auric

COCTEAU, Milhaud, and Poulenc developed their initial fascination for popular amusement independently. Cocteau and Poulenc each recall attending popular milieux during boyhood days in Paris, while Milhaud writes of early encounters that took place in 1917. Georges Auric's accounts do not refer to youthful experiences in popular venues, but convey a familiarity with a wide range of spectacles, from *parade* performances at the fairground to music-hall acts featuring American orchestras to the little circus numbers on the Boulevard St. Jacques.[1] Beginning in 1919 the four artists began sharing their interest. Together with a group of poets and painters, they met each Saturday evening and roamed the Paris streets in search of fair, circus, music-hall, and film spectacles. Their descriptions both of early experiences and of the Saturday evening wanderings allude to aesthetic principles of popular entertainment which they found appealing and useful as an inspiration for their artistic activity. The fascination of all four artists with the Parisian popular world and the interest in turning to popular spectacles as a model for their own work marked a disregard for traditional divisions separating popular and classical forms of creative expression.

POPULAR ENCOUNTERS

The abundant autobiographical writings of Poulenc, Cocteau, and Milhaud make it possible to trace each artist's exposure to popular entertainment in some detail. Francis Poulenc grew up in Paris and spent summers in the small village of Nogent-sur-Marne, just east of the capital.[2] In his interviews with the French critic Claude Rostand, Poulenc recalls that Nogent was his childhood paradise. As a young

[1] See, for example, *Le Coq* 1 (May 1920). Translations here and throughout the chapter are by the present author unless otherwise specified.

[2] Information in this paragraph from Francis Poulenc, *Entretiens avec Claude Rostand* (Paris: René Julliard, 1954), p. 17. See also Francis Poulenc and Stéphane Audel, *My Friends and Myself*, trans. James Harding (London: Dennis Dobson, 1978), p. 31.

boy, his musical taste centred around the French songwriters Henri Christiné and Vincent Scotto whose tunes were often performed by small orchestras or accordion players at dance halls known as *bals-musettes* and at popular *guinguettes* (suburban taverns) such as the Café Bebert.[3]

Beginning in 1914, when Poulenc was 15 years old, he combined summer excursions to cabarets and dancing halls in Nogent with fall and winter outings to popular entertainment spots in Paris. Poulenc recalls that he 'frequented the Parisian music-hall without stop' from age 15 to age 30. He and three of his boyhood friends particularly enjoyed the café-concerts, music-halls, and theatres in the Quartier de la République. There they attended performances by the music-hall singer Jeanne Bloch, whose songs struck Poulenc as ideal material for an opera. They also heard the *tours de chants* of Maurice Chevalier.[4] In 1914 and 1915, when Poulenc first encountered Chevalier, the music-hall singer's career was already well launched. He had performed at the Eldorado, one of the largest music-halls in Paris, and in 1909 he had started participating in song and dance acts with the popular female entertainers Mistinguett, Gaby Deslys, and Polaire at the Folies-Bergère.[5] Yet Chevalier was still a young entertainer who appeared at a wide range of popular milieux. Poulenc's personal preference lay with the *tours de chants* that Chevalier performed at modest establishments like the Petit Casino and the Carillon, where a lower-class clientele enjoyed the tradition of beer and cherry-flavoured water. The Petit Casino, dating from 1893 and spotlighting young singers, was an ideal venue for the young Chevalier.[6] In several brief statements on popular song, Poulenc conveys his admiration for the prosody of the *tours de chants* which he heard Chevalier perform at these and other haunts. The free treatment of the text and the playful cutting up of words

[3] Francis Carco mentions the presence of accordions at the *bal-musette*, in his article 'Au bal-musette', *La Danse*, Dec. 1920. Keith Daniel's statement that Poulenc was a frequent visitor at the Café Bebert in Nogent is corroborated by Poulenc's own reference in his *Correspondance* to the Bebert. See Keith W. Daniel, *Francis Poulenc: His Artistic Development and Musical Style* (Ann Arbor: UMI Research Press, 1982), p. 7; and Francis Poulenc, *Correspondance 1915–1963*, compiled by Hélene de Wendel (Paris: Éditions du Seuil, 1967), pp. 241–2.

[4] References in this paragraph to the singers whom Poulenc heard at Parisian music-halls and café-concerts come from his *Entretiens*, pp. 135–6.

[5] For a useful summary of Chevalier's career, see France Vernillat and Jacques Charpentreau, *Dictionnaire de la chanson française* (Paris: Larousse, 1968), p. 62.

[6] Information on café-concerts like the Petit Casino, the Ville Japonaise and the Carillon appears in François Caradec and Alain Weill, *Le Café-Concert* (Paris: Atelier Hachette/Massin, 1980), p. 114. Jacques Feschotte mentions the programme and the ancient tradition at the Petit Casino in *Histoire du music-hall* (Paris: Presses universitaires de France, 1965), p. 113.

were techniques carefully studied and emulated by Poulenc in many of his own songs.[7]

Poulenc's accounts of experiences in the popular arena reveal that as a musician his interests focused primarily on *guinguette, balmusette*, and music-hall. Jean Cocteau's memoirs, by contrast, express delight in a dazzling variety of performers and popular spectacles. In *Portraits-souvenir 1900–1914*, he describes childhood visits to the Nouveau Cirque where he enjoyed the aquatic pantomimes, the clown and acrobat numbers, and the spectacles inspired by American culture. Among the American acts which he recalls with special fondness were the 'Mexican marksmen', who imitated American cowboys, and the cakewalk dancers in the pantomime *Joyeux Nègres*.[8]

During Cocteau's adolescence, the Parisian music-hall absorbed his attention. The singer and dancer Mistinguett ranked highest in his esteem. Between roughly 1905 and 1908 Mistinguett was making her debut at the Eldorado as a *gommeuse* or tease, and Cocteau went there frequently to see his 'princess' perform.[9] Such outings continued well past Cocteau's adolescent years. In the fall of 1917 he attended the Casino de Paris where he saw Gaby Deslys and Harry Pilcer perform syncopated dances to an accompaniment featuring Louis Mitchell's Jazz Kings. In March of 1918 in *The Cock and the Harlequin*, Cocteau praised the extraordinary energy of the performance, pronouncing the music a 'hurricane of rhythm and beating of drums' and the dance a 'sort of domesticated catastrophe which left [Deslys and Pilcer] drunk and blinking under a battery of six floodlights'.[10]

The topic of film surfaced frequently in Cocteau's autobiographical writings. During the late 1910s, when wartime weakened the European film industry and American films assumed prominence, Cocteau became an avid fan of the Hollywood serials and the

[7] In *Entretiens*, p. 136, Poulenc acknowledged that his vocal writing in works up to *Les Mamelles de Tiresias* was influenced by Chevalier's treatment of prosody in the song 'Si fatigué'. For an analysis of Poulenc's re-creation of Chevalier's performance style, see Chapter 6.

[8] For Cocteau's account of circus outings and specifically of the Mexican marksmen and the cakewalk dancers, see *Portraits-souvenir 1900–1914*, ed. Pierre Georgel (Paris: Librairie générale française, 1977), pp. 81–91. Cocteau explains that the term 'cowboy' was not used because it was still unknown in France.

[9] Despite her role as a tease, Cocteau reverently describes Mistinguett as the 'princess of the Eldorado' whom he and his friends 'adored'. See *Portraits*, pp. 131, 134. For Mistinguett's debut at the Eldorado, see Caradec and Weill, *Le Café-Concert*, p. 107.

[10] Jean Cocteau, *The Cock and the Harlequin*, in *A Call to Order*, translated by Rollo Myers (New York: Haskell House, 1974), pp. 13–14. Translation is modified where necessary by the translation offered in Paul Collaer, *A History of Modern Music*, trans. Sally Abeles (Cleveland: The World Publishing Co., 1961), p. 224.

films of Mary Pickford and Charlie Chaplin. A description in one of Cocteau's *Carte blanche* articles of a film starring a young girl who 'swims, boxes, dances, and jumps onto moving trains, all without realizing how beautiful she is', could be a reference either to Mary Pickford or to *The Perils of Pauline*.[11] In a *Carte blanche* article of April 1919, Cocteau devoted a paragraph to Charlie Chaplin, pronouncing him the 'modern puppet' who 'addresses all ages and all peoples', and revelling in the light-hearted spirit of a chase scene in Chaplin's film *Sous les armes*. References in *Portraits* to three Lumières films—*The Sprinkled Sprinkler, Beer*, and *Babies on the Shore of the Water*—indicate that Cocteau was also familiar with the earliest film projections shown in Paris.

The autobiographical writings of Darius Milhaud, like those of Cocteau, reflect an absorption with French and American popular entertainment and, in addition, an interest in the popular tunes and rhythms of South America. Although Milhaud provides less information than Poulenc and Cocteau about childhood experiences in the popular realm, his descriptions of contact in the late 1910s and early 1920s are quite detailed. From the autobiography, *Notes without Music*, we learn that his first important experience occurred in 1917, when the French poet and diplomat Paul Claudel was appointed Ambassador to Brazil and Milhaud accompanied him as secretary.[12]

Milhaud and Claudel arrived in Rio de Janeiro on 1 February 1917, in the midst of the six-week Carnaval. An atmosphere of 'crazy gaiety' reigned, with people singing and dancing in the streets, and Milhaud found himself immersed in a new musical world. Two aspects of the Brazilian Carnaval music made a particular impression on him: first, the tone quality of instruments like the *violau*, a guitar, and the *choucalha*, a percussion instrument containing iron filings and a rod at one end which rotated; second, the syncopated rhythms of the popular dance tunes. Of the latter, Milhaud wrote: 'There was an imperceptible pause in the syncopation, a careless catch in the breath, a slight hiatus which I found very difficult to grasp. So I bought a lot of maxixes and tangos, and

[11] *Carte blanche*, in *Le Rappel à l'ordre* (Paris: Éditions Stock, 1948), p. 126. Richard Axsom writes that Mary Pickford was one of Cocteau's favourite film stars. According to Leonide Massine, Pickford was Cocteau's model for the Little American Girl in *Parade*. Axsom, '*Parade*': *Cubism as Theater* (New York: Garland Publishing, 1979), p. 42. Subsequent references in this paragraph to Charlie Chaplin are taken from *Carte blanche*, p. 93; to the Lumières films, from *Portraits*, p. 68.

[12] *Notes without Music*, trans. Donald Evans, ed. Rollo Myers (London: Dennis Dobson, 1952), pp. 63–4.

tried to play them [i.e. on the piano] with their syncopated rhythms that run from one hand to the other.'[13]

In November of 1918, Claudel was sent from Rio to Washington DC to represent France on an inter-allied economic mission. Once again Milhaud accompanied him. Interestingly, Milhaud's memoirs make no mention of the American popular music which he may have heard in the United States. Instead he describes the assortment of musics of different nationalities which filled his ears in Puerto Rico, where he and Claudel stopped en route to America.

From Washington DC Claudel and Milhaud returned to Paris in early 1919, and Milhaud pursued his interest in popular entertainment there. His early book of memoirs, *Études*, describes how he attended satirical skits performed by the famous clown Grock and by the Fratellini clowns at the Cirque Médrano. Milhaud assigns no date to these outings but since Grock made his debut at the Cirque Médrano in 1904 and the Fratellinis around 1915, it seems likely that Milhaud began frequenting the circus before his trip to Brazil and continued upon his return.[14]

Études also contains portraits of other favourite popular haunts and entertainers, again without reference to the period during which Milhaud attended. In one brief sketch, Milhaud describes a music-hall in the provinces outside Paris where the singer, Damia, starred in a revue called *La Fille Elisa*. Milhaud writes that he found the revue 'detestable', but had great admiration for Damia. In another vignette, he recalls spending evenings at a bar called L'Ours on rue Daunou, where Parisians danced to the accompaniment of a musical saw played by a M. Andolfi, and to an orchestral accompaniment of 'blues' or 'rags'.[15]

In 1917 when Gaby Deslys and Harry Pilcer performed their syncopated dance at the Casino de Paris, Milhaud was in Brazil. He undoubtedly heard about Mitchell's Jazz Kings, however, for in a 1924 lecture at the Sorbonne entitled 'Les Ressources nouvelles de la musique (jazz-band et instruments mécaniques)', he credited Deslys and Pilcer with having introduced the 'jazz band' to

[13] *Notes without Music*, p, 64. The maxixe was a Brazilian urban popular dance that appeared in Rio de Janeiro around 1870. It soon became synonymous with the Brazilian tango.

[14] For the dates of Grock's debut at the Cirque Médrano, and of the Fratellinis' early stardom at the Médrano, see Adrian, *Histoire illustrée des cirques parisiens d'hier et d'aujourd'hui* (Paris: Adrian, 1957), pp. 113, 117.

[15] For the preceding discussion of popular entertainers Milhaud came to know in Paris, see *Études*, pp. 79–85. 'Blues' is probably not an appropriate term for the musical accompaniment played by M. Andolfi on the saw, unless Milhaud heard these performances in the early 1920s.

France.[16] The same lecture also contained a detailed description of the drum technique of a 'M. Buddy' who played with the 'Syncopated Orchestra'. The reference may have been to performances by Will Marion Cook's Southern Syncopated Orchestra in October 1919; if so, then such performances comprised Milhaud's first exposure to American ragtime and syncopated dance music.

A second encounter took place in 1920 when Milhaud was in London for productions of his ballet, *Le Bœuf sur le toit* (The Ox on the Roof). There he heard the white American musician Billy Arnold and his orchestra, the Novelty Jazz Band, perform in a Hammersmith dance-hall.[17] The varied timbres, the syncopated rhythms, and the precision of Arnold's music greatly impressed Milhaud and offered, in his view, a refreshing contrast to the 'crude sounds' of the *bals-musette* orchestras in Paris. In *Notes without Music*, he observed: 'The new music was extremely subtle in its use of timbre . . . The constant use of syncopation in the melody was of such contrapuntal freedom that it gave the impression of unregulated improvisation, whereas in actual fact it was elaborately rehearsed daily, down to the last detail.'[18]

Milhaud's enthusiasm for the exotic sounds of American music, and his reference to *bals-musettes* orchestras as 'crude', reveal a tendency to take special delight in popular idioms of foreign cultures. Yet his general receptivity to popular entertainment of all kinds made him an avid leader and participant in Saturday evening excursions to Parisian fair, circus, cinema, and music-hall. Sometime in 1919, and probably at least a year before Milhaud heard Billy Arnold perform in London, he, Poulenc, Cocteau, and Auric, along with Durey, Honegger, and Tailleferre, met every week at Milhaud's apartment in Montmartre.[19] They were joined by the

[16] See Darius Milhaud, 'Les Ressources nouvelles de la musique (jazz-band et instruments mécaniques)', *L'Esprit nouveau* 25 (1924): n.p. Milhaud may well have heard the Jazz Kings perform at the Casino de Paris in the 1920s. It is important that his use of the term 'jazz' be placed in quotes, since the music he heard was actually a form of ragtime and syncopated dance band music.

[17] See the account by Milhaud, 'La Musique "pas sérieuse" ', *Jazz* 325 (Jan. 1984), p. 55.

[18] *Notes without Music*, p. 102.

[19] According to Georges Auric, it was Milhaud who proposed the Saturday evening meetings. See Georges Auric, *Quand j'étais là* (Paris: Bernard Grasset, 1979), p. 136. The date of 1919 cannot be absolutely verified since there are conflicting reports. In *Notes without Music* (p. 83), Milhaud states that the meetings did not begin until Collet's article on Les Six appeared on 23 January 1920. However according to Frédéric Robert, the biographer of Louis Durey, Collet baptized the Six after attending one of their meetings. Robert's view is corroborated by Willy Tappolet, biographer of Honegger, and by Keith Daniel. See Robert, *Louis Durey: L'aîné des Six* (Paris: Les Éditeurs français réunis, 1968), p. 29; Willy Tappolet, *Arthur Honegger* (Zurich: Atlantis Verlag, 1954), p. 37; and Keith Daniel, *Francis Poulenc*, p. 18. Since the meetings cannot have predated Milhaud's return from Brazil, it seems most likely that they began in 1919.

painters Valentine Gross, Marie Laurencin, Irène Lagut, and Guy-Pierre Fauconnet, the poets Raymond Radiguet, Lucien Daudet, and Jean Hugo, the pianist Marcelle Meyer, and numerous other artists. Although the order of events varied, the customary pattern was to move from Milhaud's apartment to a small restaurant such as the Petit Bessonneau on rue Blanche, and then to the Montmartre fair, the Cirque Médrano, a music-hall like the Folies-Bergère, or a cinema showing films of Charlie Chaplin. The evening usually concluded at Milhaud's flat, with poetry readings and performances of recent compositions.[20]

During the two years which followed, as the Saturday evening meetings became an institution, Milhaud often urged that the reunions be transferred from his apartment to a bar or bistro which the group could call its own. Sometime in 1921 the composer and pianist Jean Wiéner, whom Milhaud had known since 1911 when they both studied at the Conservatoire, responded to Milhaud's suggestion with a proposal for a wine store on rue Duphot. Cocteau joined the two men in making arrangements with the French owner, Louis Moyses. A bar was set up in one room with a restaurant in the other, the walls were decorated with posters bearing the names of the guests, and in late February of 1921 the bar Gaya, christened after a Spanish wine which Moyses used to sell there, became the new meeting and dining place of the Saturday group.[21]

From its inception the Gaya promoted a Parisian taste for American popular music. The decision to feature American music was made by Wiéner. In his memoirs, Wiéner explains that at this time his knowledge of American dance music stemmed principally from two events: first, a study of several measures of ragtime which his friend Yves Nat had notated on an envelope and brought back from the United States to show him; second, his attendance at the ragtime dance performance of Deslys and Pilcer.[22] Not long before

---

[20] The number of artists who participated in the Saturday excursions clearly varied, for lists of names provided by Georges Auric, Francis Steegmuller, and others do not always coincide. The account in this paragraph assembles the artists mentioned both by Auric and by Steegmuller. See Auric, *Quand j'étais là*, pp. 140–1, and Steegmuller, *Cocteau: A Biography* (Boston: Little, Brown, & Co., 1970), p. 246. For a discussion of the restaurants and the popular institutions most commonly frequented, see Milhaud, *Notes without Music*, pp. 83–4.

[21] For an account of the founding of the bar Gaya, see Jean Wiéner, *Allegro appassionato* (Paris: Pierre Belfond, 1978), p. 43. According to Milhaud, Auric, and Cocteau, the Gaya was a nightclub where Wiéner had been playing piano for some time before the Saturday evening meetings began. However, since Wiéner himself records the story differently, it seems advisable to rely on his account. The bar's décor, the date of its opening and its function as a meeting and dining place are based on Steegmuller's account in *Cocteau: A Biography*, p. 263. Steegmuller draws his information from a diary which the artist Jean Hugo kept during the 1920s.

[22] Wiéner, *Allegro appassionato*, p. 28.

the opening of the Gaya, moreover, Wiéner reports that he 'received some tunes' by George Gershwin, Vincent Youmans, and Fletcher Henderson, presumably in the form of sheet music or recordings.[23] As a pianist, he probably experimented with his own keyboard renditions of this music, just as he may have tried his hand at the American tunes which were performed so frequently in Parisian music-halls and public squares.

The Gershwin and Youmans tunes provided the chief musical entertainment on the Gaya's opening night. Wiéner played them on an upright piano which he had rented, while Milhaud, Cocteau, and Marcelle Meyer furnished the percussion section, playing a bass drum and kettledrum borrowed from Stravinsky. A black American saxophonist and banjoist named Vance Lowry, who had performed with Louis Mitchell in London in 1915, played solos and joined Wiéner for duets.[24] This first concert established the Gaya as a meeting place where artists, poets, and composers in Paris could enjoy performances of American Tin Pan Alley, blues, and dance tunes. A marvellously detailed paragraph, which Wiéner tells us he discovered in one of his notebooks from the 1920s, recreates the Gaya's lively social environment. He describes how at one table Diaghilev sat with Picasso, Misia Sert, and the Russian poet Boris Kochno; in another corner Satie spoke with René Clair and Jane Bathori; and at the back of the bar Georges Auric joked with the American singer Yvonne George. The presence of Maurice Chevalier and Mistinguett as well must have delighted Francis Poulenc who was also a regular visitor. Wiéner observes that among these and other artists and intellectuals who frequented Moyses's establishment, several knew the Wiéner–Lowry repertoire extremely well and had distinct preferences for certain American tunes. The painter Fernand Legér, for example, often requested a performance of W. C. Handy's *St Louis Blues*, which was a standard for white dance bands in the United States. *Old Fashioned Love*, a jazz tune by James P. Johnson, was a favourite of Poulenc who 'listened religiously, leaning on the piano', as Cocteau accompanied Wiéner on the drums and cymbals.

[23] It is unlikely that tunes of Fletcher Henderson were made available to Wiéner, since Henderson did not start recording until 1923, and sheet music arrangements of his tunes could only have appeared after that. Wiéner is quite likely to have had access to Gershwin's revue tunes of the early 1920s and Youmans's early Tin Pan Alley songs.
[24] Information on the Gaya's opening night is gleaned from Wiéner, *Allegro appassionato*, p. 43 and from Cocteau, *La Jeunesse et la scandale*, in *Œuvres complètes* (Paris: Marguerat, 1950), ix. 330. Information on Vance Lowry's London engagement with Louis Mitchell's band, the Seven Spades, appears in Robert Goffin, *Jazz: From the Congo to the Metropolitan* (New York: Da Capo Press, 1975), p. 58.

Wiéner applied his interest in popular music both to evenings at the Gaya and to the arrangement of concerts in Paris. His adventurous programming was responsible for one of the first concerts to include an American dance band, Billy Arnold's, alongside a Pleyela piano performance of *The Rite of Spring* and a performance of Milhaud's Sonata for piano and winds.[25] The presentation of this event at an old, well-established concert hall, the Salle des Agriculteurs, reveals how extensively the traditional boundaries between art and popular music were being challenged.

As the enthusiasm of Parisian artists and intellectuals for American popular music mounted, it became necessary for the Gaya's clientele to find larger quarters. On 10 January 1922, Louis Moyses left the rue Duphot and set up a new bar on the rue Boissy d'Anglas. This establishment adopted the name of Milhaud's ballet, *Le Bœuf sur le toit*, and within a short time it surpassed the Gaya in popularity. Several comments by critics and artists attest to the significant role Le Bœuf sur le toit played in Parisian musical and social life. The American critic Edmund Wilson, for example, pronounced Le Bœuf one of the nerve centres of the musical revolution in France.[26] Cocteau observed that Le Bœuf was 'not a bar at all, but a kind of club, the meeting place of all the best people in Paris, from all spheres of life—the prettiest women, poets, musicians, businessmen, publishers—everybody met everybody at the Bœuf'.[27]

During the 1920s the Parisian obsession with American popular music instilled in Darius Milhaud a desire to learn more about early jazz. In 1922 when the opportunity arose to visit the United States, he accepted. The American pianist E. Robert Schmitz, an ex-piano-pupil of Debussy's who headed the Pro Musica Society and took a strong interest in promoting the performance of French music in America, arranged for Milhaud to come and conduct and perform his works.[28] Milhaud travelled to Princeton, Vassar, and Harvard, conducted the Philadelphia Orchestra, and played the piano with the New York City Symphony Orchestra. His account of the trip in *Notes without Music* portrays his experiences with American

[25] Descriptions of the Gaya's social atmosphere taken from *Allegro appassionato*, p. 44. Information on this *concert salade* comes from a concert programme of December 1921 included in *Allegro appassionato*. Similarly eclectic programmes were later featured at the bar, Le Bœuf sur le toit.

[26] Edmund Wilson, 'The Aesthetic Upheaval in France: The Influence of Jazz in Paris and Americanization of French Literature and Art', *Vanity Fair* 17 (Feb. 1922).

[27] Steegmuller, *Cocteau*, p. 281.

[28] Through his Pro Musica Society, a chamber music organization with chapters all over the United States, Schmitz later arranged Ravel's 1928 American tour.

orchestras, but gives special attention to the popular styles and particularly to the black music he encountered in the United States, for it was Milhaud's desire to 'find out all I could about negro music'.[29] Milhaud's serious interest in early jazz astonished American reporters accustomed to the indifference American composers showed towards the new style. During his visit, newspapers in the United States displayed such headlines as 'Milhaud Admires Jazz' and 'Jazz Dictates the Future of European Music'.

Milhaud's introduction to black musical styles took place in New York. He describes a meeting with the composer, arranger, and concert performer Harry T. Burleigh who 'played me negro folktunes and hymns which interested me keenly'.[30] In his *Études*, Milhaud compared the mood of black spirituals, which he came to know principally through Burleigh's arrangements, with the sombre tone of popular numbers like *St Louis Blues* and *Aunt Hagar's Children's Blues*. 'In blues', he noted, 'there is the same tenderness, the same sadness as that [sorrow] which inspired the slaves . . .'[31]

Milhaud was the first European composer of the early twentieth century to come into contact with Harlem blues. Although his narrative in *Notes without Music* does not specify which bar or club he attended, in *Études* he makes passing reference to a 'dance hall' called the Capitol on Lenox Avenue near 140th St. in Harlem. His description of the black female singer who performed at this club matches the description, in *Notes without Music*, of the vocalist he heard in Harlem. One can speculate, then, that the Capitol was the place where Milhaud first encountered Harlem blues performed by any one of a number of female singers—Edith Wilson, Leona Williams, Mamie Smith, Ethel Waters—who were recording on OKeh, Columbia, Aeolian, and other American labels in 1922. Milhaud's wife Madeleine thinks it likely that the singer whom Milhaud so admired at the Capitol was Ethel Waters.[32] His

[29] *Notes without Music*, pp. 114–18.

[30] For information on Harry T. Burleigh, see Eileen Southern, *The Music of Black Americans*, 2nd edn. (New York: W. W. Norton, 1983), pp. 267–8.

[31] *Études*, p. 57. Although *St Louis Blues* does not follow blues form, the melody may have struck Milhaud as similar to other blues melodies and thus evocative of spirituals as well. According to Gunther Schuller, Milhaud brought an Original Memphis Five recording of *Aunt Hagar's Children* back with him to Paris. See Schuller, *Early Jazz: Its Roots and Musical Development* (New York: Oxford University Press, 1968), p. 186.

[32] The author is grateful to Madeleine Milhaud for this insight provided in a letter of 5 June 1989. The full name of the club was the Capitol Palace, located on Lenox Avenue between 139th and 140th Streets. See Leroy Ostransky, *Jazz City* (Englewood Cliffs, NJ: Prentice Hall, Inc., 1978), p. 200. Samuel Charters and Leonard Kunstadt cite the female blues singers recording on American labels in the early 1920s. See Charters and Kunstadt, *Jazz: A History of the New York Scene* (Garden City, NY: Doubleday & Co., 1962), pp. 95–105.

portrayal of her performance conveys his enthusiasm for an emotionally charged sound world: 'Against the beat of the drums the melodic lines criss-crossed in a breathless pattern of broken and twisted patterns . . . With despairing pathos and dramatic feeling, [a negress] sang over and over again, to the point of exhaustion, the same refrain to which the constantly changing melodic pattern of the orchestra wove a kaleidoscope background.'[33]

The performance at the Capitol inspired Milhaud to purchase some blues recordings, and before leaving New York he bought two—*I wish I could shimmy like my sister Kate* and *The wicked five blues*—on a label called Black Swan. The Black-owned company emerged in 1921 to record the current black music, both jazz and blues; thus it is quite possible that Milhaud obtained a recording by the female singer whom he heard at the Capitol.[34]

In *Notes without Music*, Milhaud also mentions that he attended 'negro theatres' and 'dance halls' in New York. The reference to black theatres, coupled with the description in *Études* of Noble Sissle's and Eubie Blake's musical *Shuffle Along*, indicates that within the realm of black musical theatre Milhaud became acquainted with one of the most important American shows of the early twentieth century. Written, performed, produced, and directed by blacks, *Shuffle Along* was the first black musical to appear in white theatres throughout the United States. Since it opened in New York on 23 May 1921 and ran for fourteen months before setting out on tours of Boston, Chicago and twelve other American cities, Milhaud may have seen a performance at the 63rd Street Theatre on Broadway. His discussions of American 'jazz' place *Shuffle Along* and the musical *Liza* by Maceo Pinkard in the category of 'dance and theatre music' written in a 'jazz negro style'. Although 'jazz' is not an appropriate term for the black theatre music which Milhaud heard in New York, he was probably referring to the ragtime music and the jazz dances in *Shuffle Along*. The dances included carryovers from minstrelsy and vaudeville, soft shoe, tap dances and, by 1924 in some of the later black musicals, the Charleston.[35]

Milhaud's reference to a 'jazz negro style' reveals his awareness

---

[33] *Notes without Music*, p. 118.

[34] At the Institute of Jazz Studies, Rutgers University, I heard Black Swan recordings featuring Henderson's Dance Orchestra. This orchestra may have appeared on the recordings purchased by Milhaud.

[35] The importance of *Shuffle Along* and the use of ragtime music and jazz dancing are discussed in Robert Kimball's liner notes to *Sissle and Blake's 'Shuffle Along'* (New World Records 260), p. 1. Eileen Southern discusses the impact of *Shuffle Along* on black musicals of the 1920s and 1930s and identifies the 'jazz' dances in *The Music of Black Americans*, pp. 428–9, 432.

of the difference between black 'jazz', and white 'jazz' or what he termed 'American jazz'. In his *Études*, he remarks that in the music of black Americans, 'the primitive African side has remained more firmly rooted . . . and it is there that one must see the source of this terrific rhythmic power'. Milhaud's admiration for the 'primitive' and his association of the 'primitive' with the exotic coincided with Blaise Cendrars's vision of the noble savage and inspired their collaboration in 1923 on the blues ballet *La Création du monde*.

The appeal for Milhaud of the improvisatory, highly rhythmic performance style of blues bands and black theatre orchestras led him to conclude that in comparison, white dance orchestras like Paul Whiteman's 'had the precision of an elegant, well-oiled machine, a sort of Rolls Royce of dance music . . .'.[36] Yet when he identified specific musical traits of black 'jazz', Milhaud tended to confuse it with white 'jazz'. In his *Études*, for example, he praises several features of black performances—simultaneous use of major and minor chords, and quarter tones produced by a mixture of glissando and vibrato—which are also admired in his general discussion of white 'jazz'.[37] It seems, then, that Milhaud's encounter with black musical styles in New York awakened him to the subtleties of American popular music and to the subcategories within the 'jazz' heading, while increasing his difficulty in formulating a clear definition of the jazz idiom. The lure of 'primitive' Black American styles made objective musical distinctions more difficult.

THE APPEAL OF POPULAR ENTERTAINMENT

Statements by Poulenc, Auric, and Milhaud addressing the appeal of popular entertainment were not confined to a particular decade, but, rather, spanned a number of years. Some of the writings, which took the form of articles embracing a new cause, appeared as early as 1920 when the three composers were participating in their group excursions to popular establishments. Other accounts, in which the composers looked back on the years 1900–24 and on childhood experiences and recalled their fascination with the popular world, were published as late as 1979 in the context of an autobiography or interview.[38] Yet taken together, the composers' writings reveal a

[36] *Notes without Music*, p. 117.

[37] Distinctions between 'American jazz' and 'Negro jazz' and the importance of improvization in Negro jazz are discussed in *Études*, pp. 55–7. Overlaps between black and white 'jazz' elements are made on p. 20.

[38] Auric's autobiography, *Quand j'étais là*, was published in 1979. Poulenc's *Entretiens* appeared in 1954 and his *My Friends and Myself* in 1963. Milhaud's *Notes without Music* appeared in 1952.

remarkable consistency in point of view; features praised in the 1920s tended to be admired again in the 1950s and 1970s. Many of the aesthetic principles discussed by the three composers, moreover, were first introduced by Jean Cocteau in his manifesto of 1918, *The Cock and the Harlequin* and in his *Carte blanche* articles of 1919. Cocteau's influence on the attitudes of Poulenc, Auric, and Milhaud can best be understood by identifying the aspects of popular entertainment which they found attractive, and then turning to the early writings of the man who considered himself their spokesman.

For Francis Poulenc, the world of popular amusement evoked his past, and specifically the everyday sights and sounds of his youth. Nostalgic memories of experiences on the streets of Nogent-sur-Marne and in the café-concerts and music-halls of Paris were an important component of Poulenc's adult sensibility; he came to view both the tunes of his favourite songwriters, Christiné and Scotto, and the ambiance of popular milieux as his 'folklore'.[39] Poulenc's was very much a French folklore, moreover, since it was rooted in the popular musics and localities of France. In a statement written for the *Bulletin de la Phonothèque Nationale* in 1963, Poulenc described the inflection of everyday French street life and street music which permeated his art compositions and traced it to summers in Nogent: 'All my first compositions, and indeed everything one considers my amorous side, my erotic side, comes from Nogent-sur-Marne, and from this kind of stale smell of fried potatoes, of dinghies, and of the blare of the distant band.'[40] The references here to a 'stale smell' and to distant sounds of a band suggest a wistful component of street life which did not escape Poulenc's notice. Indeed Poulenc's nostalgic view and his memories of the past were coloured by feelings of melancholy and an awareness of the sadness beneath surface gaiety. Such feelings played an important role in Poulenc's attraction to the popular realm. In a statement on Apollinaire, he described the presence in Apollinaire's poetry of a 'style of a popular sentimental ballad' and a 'sympathy for the suburbs and the little boats on the Marne', and remarked upon the poetic charm of this bittersweet tinge.[41]

---

[39] Poulenc, *Entretiens*, p. 17.

[40] Quoted by Keith Daniel in *Francis Poulenc*, p. 7. The statement is taken from 'Hommage à Francis Poulenc', *Bulletin de la Phonothèque Nationale*, special supplement to No. 1 (Jan.–Mar. 1963): 14.

[41] Keith Daniel translates Poulenc's entire statement in *Francis Poulenc*, p. 32. The observation appears in Nino Franck, 'Poulenc à Montmartre', *Candide*, 28 Apr. 1932, n.p. For other nostalgic, melancholy memories of popular institutions in Nogent, see Poulenc, *Correspondance*, pp. 241–2.

Poulenc's childhood exposure to a variety of popular institutions, coupled with his interest from a very early age in both popular and concert music styles, instilled in the young composer a love for musical diversity. In his conversations with Stéphane Audel, Poulenc alluded to the piano music which he heard and played at home and to the music-hall songs which he enjoyed at popular milieux near the Bastille, and observed that he found nothing jarring in this diverse blend of musical idioms:[42]

I've often been reproached about my 'street music' side. Its genuineness has been suspected and yet there's nothing more genuine in me. Our two families ran their business houses in the Marais district, full of lovely old houses, a few yards from the Bastille. From childhood onwards I've associated café tunes with the Couperin suites in a common love without distinguishing between them.

It was consistent with this point of view that Poulenc took great pride in the popular tunes he created. When Manuel Rosenthal asked Ravel for his opinion of Poulenc's music, Ravel replied with a comment which apparently delighted the young composer: 'What I like is his ability to invent popular tunes . . .'[43]

The common love for café tunes and Couperin suites made Poulenc receptive to a range of musical influences. In a letter to Paul Landormy in which he discussed the composers who had left their mark on his style, Poulenc named such popular Parisian music-hall songwriters as Mayol, along with countless French, German, and Russian art composers.[44] Significantly, the German composers mentioned were eighteenth- rather than nineteenth-century musicians. Poulenc did not cite Wagner, for instance, and consciously shunned Wagner's influence because he wished French music to return to classical ideals of 'simplification', 'counterpoint', 'melody', and 'precision'.[45] Although he never drew direct parallels between the fair or the music-hall and simplicity, his frequent use of terms like 'coarse', 'clear', and 'robust' in discussions both of popular music and of a new simple French art music suggests that he considered the French

---

[42] Poulenc and Audel, *My Friends and Myself*, p. 31. Poulenc's mother was a pianist and introduced him to Mozart, Schumann, and Chopin.
[43] Manuel Rosenthal, *Satie, Ravel, Poulenc* (Madras, New York: Hanuman Books, 1987), pp. 74–5.
[44] Paul Landormy, *La Musique française après Debussy* (Paris: Gallimard, 1943), p. 162. Poulenc probably wrote the letter not long after the 1917 production of *Parade*.
[45] Poulenc invoked these classical ideals when he identified the common goal of Les Six. See Landormy, *La Musique française après Debussy*, p. 116. Strangely, Poulenc never mentioned eighteenth-century French composers in his discussions of precision and simplicity.

popular tradition an important source of inspiration for the revival of a classical simplicity in music.

In a brief article entitled 'Accent populaire' which Poulenc published in the broadsheet *Le Coq* in November of 1920, he defended French popular melody as an effective counterweight to the dense, chromatic writing of nineteenth-century German composers and their French followers. Popular tunes, he argued, imparted not only a directness of expression but a national voice to modern music:[46] 'Vulgar melody is good in that it is original. I like Romeo, Faust, Manon, and even Mayol's songs. Refinement [German and Debussyian] obliterates the folk accent in your "modern music". When refinement and that folk accent combine in a nation (as in Russia), that nation has found its music.' In the same year, Poulenc explained to Paul Collaer that his songs *Cocardes*, inspired by the *tour de chants* of Parisian café-concert and music-hall, were 'coarse' pieces that were 'stripped of artifice' and evoked an atmosphere of Paris.[47]

The modern French art composers whom Poulenc admired most were those who incorporated the simplicity of French popular tunes. Satie, for instance, was upheld by Poulenc as a model of French clarity. In his letter to Paul Landormy cited above, Poulenc disparaged Debussyism and the Impressionism of Ravel and Schmitt, and praised Satie's music because it was 'clear', 'robust', and 'as frankly French as Stravinsky's is Slavic'. Poulenc also celebrated Satie's ballet *Parade* for its use of popular sounds which defied the academic and the sublime. In an account of the scandal provoked by *Parade*, Poulenc expressed delight in the dancing of a one-step in the ballet, and in the unprecedented manner in which music-hall entertainment 'invaded art with a capital A'.[48]

Georges Auric was equally fervent in his support for the simple, down-to-earth nature of popular entertainment and for the usefulness of popular spectacles in diverting French art composers from the refined Wagnerian world. An article written by Auric for *Le Coq* in June 1920 contained the following protest: 'Why begrudge us the circus, the music-hall, the fair of Montmartre? . . . We needed this raw, crisp uproar of sound. Too bad if it disperses the heavy seductions of Debussyism a bit too explosively.'[49] Auric went on to

---

[46] Paul Collaer translates Poulenc's statement in *A History of Modern Music*, p. 224. Romeo was a Cuban composer of popular music whom Poulenc probably came to know through Milhaud when the latter returned from his trip to Brazil.

[47] Paul Collaer, 'I "Sei": studio del'evoluzione della musica francese dal 1917 al 1924', *L'Approdo musicale* 19–20 (1965): 45.

[48] Poulenc and Audel, *My Friends and Myself*, p. 68.

[49] Auric, 'Après la pluie le beau temps', *Le Coq* 2 (June 1920).

praise the jazz band because its 'counterpoint of noises, rhythms, and shouts', superimposed on a simple everyday dance tune, disturbed people and roused them from their torpor. His choice of the term 'counterpoint', with its suggestion of classicism, paralleled Poulenc's use of classical terms to describe the simplicity of the popular realm. Like Poulenc, moreover, Auric applauded Satie's music for incorporating the simplicity of jazz and the circus in order to effectively counteract the 'clouds and sirens' of Debussyism.[50]

Auric's association of the popular world with an 'impiety' and humour often prompted him to speak of popular spectacles as a tool for mocking the serious concert musician.[51] In his autobiography, *Quand j'étais là*, he described a symphonic competition sponsored by Vincent d'Indy and Alfred Bruneau, in which he and Milhaud performed *Cinéma-Fantasie*, a concert piece written by Milhaud in imitation of the musical accompaniments of Charlie Chaplin films. Auric and Milhaud purposely selected this piece because they wished to poke fun at the sponsors and disrupt the serious tone of the competition. For Auric, the appeal of such light-hearted pieces as *Cinéma-Fantasie* lay in engendering a disrespect for the traditional notion of the composer as 'genius'. In an article for *La Nouvelle Revue française*, he explained that acrobatic acts and other popular spectacles pleased him because they could be judged not on the basis of intelligence, but on the basis of success. The acrobat was successful, Auric observed, if he landed on his feet. 'To be moved or to be insensitive is fine', he declared, 'but, in my view, one has been too intelligent'.[52]

Like Poulenc, Auric readily merged popular and concert music arenas and viewed the myriad sounds of both as components of a great entertainment world. In *Quand j'étais là* he recalls discovering American films, circus events, and fairground tunes with his fellow artists, at the same time that he was attending 'serious' concert music events. The value of this diversity for French art composers was clearly felt by Auric and acknowledged in an article on contemporary music which appeared in *Les Écrits nouveaux* in 1922.

[50] Georges Auric, 'Les Ballets-Russes: à propos de *Parade*', *La Nouvelle Revue française* 16 (1921): 224.

[51] In his 1978 preface to *Le Coq et l'arlequin*, Georges Auric speaks of the 'impiety' of Les Six in opposing Wagner. See 'Preface' in Jean Cocteau, *Le Coq et l'arlequin* (Paris: Éditions Stock, 1979), p. 29.

[52] For Auric's account of his performance of *Cinéma-Fantasie* with Milhaud, see *Quand j'étais là*, pp. 155–6. His attack on the notion of 'genius' appears in Auric, 'Les Ballets-Russes: à propos de *Parade*', p. 226. Paul Collaer does not include *Cinéma-Fantasie* in the work list compiled in his biography of Milhaud, and Milhaud never mentions a piece with this title. *Cinéma-Fantasie* was apparently an early version of the work which came to be called *Le Bœuf sur le toit*.

Here he expressed the need for an alternative to the 'charming mystery' of Debussy's writing and described the different styles which he and other post-Debussyian composers (Poulenc, Milhaud) were exploring in their music: 'At this point, one can thus mix and mingle many extremes: refrains of Mayol, waltzes of Chopin, certain tunes heard in the evening in small country cafés, tunes which the mechanical pianos do not tire of rolling out, so many pages of Mozart, which are sweeter than the sweetest caresses . . .'[53]

Popular entertainment had a serious as well as a light-hearted side, and Auric found both appealing. Like Francis Poulenc, he associated Parisian streets and popular institutions with a melancholy nostalgia and believed that beneath the noise and bustle there lurked a poetry, or what he termed a 'nightingale's song'.[54] The concert pieces which Auric applauded—and Satie's *Parade* was prominent among them—were those which used the sounds of Parisian popular milieux to create a bittersweet inflection and a wistful yearning for the past. In his preface to the piano version of *Parade*, Auric praised Satie's music for refusing to seduce the listener with brilliance and for offering instead a 'sadness of trestles' and a 'nostalgia of a barrel organ which will never play Bach fugues'. In his review of *Parade* for *La Nouvelle Revue française*, he called the ballet a 'touching nostalgia of trombones and drums, boulevard Saint Jacques or boulevard Pasteur'.[55] The emphasis on Parisian localities echoed Poulenc's special concern with French sights and sounds as well as Poulenc's admiration for the French voice in Satie's music.

Darius Milhaud identified and praised many of the same principles of popular entertainment which captivated Poulenc and Auric. He shared their receptivity to an eclectic assortment of sounds, their fondness for the straightforward, irreverent nature of popular spectacles, and their interest in simplicity. In his descriptions of formative musical experiences, Milhaud alluded frequently to the values of diversity and simultaneity. His account of the trip with Claudel in 1918, for example, portrays in glowing terms the musical simultaneity of Puerto Rico, where American influences merged with traces of Spanish colonization. Milhaud was particularly struck by the presence, at any given moment, of Spanish couples dancing

[53] Georges Auric, 'La Musique: quelques maîtres contemporains', *Les Écrits nouveaux* 9 (Mar. 1922): 74.

[54] Georges Auric, Preface to the Four-Hand Piano Version of Erik Satie's *Parade* (Paris: Rouart-Lerolle, 1917).

[55] Auric, 'Les Ballets-Russes: à propos de *Parade*', p. 225.

the tango, an American military band performing Sousa marches and foxtrots, and the Cuban composer Romeo playing dances in which 'Bach-like allegro themes' were combined with the 'sharp, syncopated rhythms of popular music' and the 'emphatic grinding rhythm of the *guitcharo*'.[56]

For Milhaud's young ears, this juxtaposition of musics from different nationalities and the coexistence of popular and classical idioms within a single dance piece was fascinating and exotic rather than discordant.[57] Milhaud welcomed such an array of hetero-geneous musical elements as an important source of inspiration for the art composer. In his account of the Saturday evening visits to Parisian fair, circus, cinema, and music-hall, he particularly em-phasized the range of entertainment venues, the simultaneous playing of different music-hall and revue tunes, and the significance of these evening excursions for his musical writing and for that of his fellow musicians:[58]

After dinner, lured by the steam-driven roundabouts, the mysterious booths, the 'Daughter of Mars', the shooting galleries, the games of chance, the menageries, the din of the mechanical organs with their perforated rolls seeming to grind out simultaneously and implacably all the blaring tunes from the music-halls and revues, we would visit the Fair of Montmartre, or occasionally the Cirque Médrano, to see the Fratellinis in their sketches, so steeped in poetry and imagination that they were worthy of the Commedia dell'Arte. We finished up the evening at my house. The poets would read their poems and we would play our latest compositions. Some of them, such as Auric's *Adieu, New York*, Poulenc's *Cocardes*, and my *Bœuf sur le toit*, were continually being played . . . Out of these meetings, in which a spirit of carefree gaiety reigned, many a fruitful collaboration was to be born; they also determined the character of several works strongly marked by the influence of the music hall.

Milhaud's fascination with the simultaneous barrage of popular sounds and his fondness for mixing popular and art music worlds made him a great admirer of Jean Wiéner. In a discussion of the musical events organized by Wiéner at the Bar Gaya and at Parisian concert halls, Milhaud praised Wiéner's open-minded approach to concert-programmming and his willingness to find value in a

[56] *Notes without Music*, p. 77. By *guitcharo*, Milhaud probably meant the *guiro*, a scraper found in Brazil and other regions of the Americas.

[57] Milhaud's writings contain numerous allusions to the exoticism of foreign popular musics, especially South American and black American. In his *Études*, for example, he describes the presence in black American melodies of a 'lyricism which only oppressed races could produce'. He also speaks of the 'savage African character' of black music in America. *Études*, pp. 56, 59.

[58] *Notes without Music*, pp. 83–4.

remarkable variety of musical styles. He was impressed by Wiéner's conviction that programmes need not be devoted exclusively to performances by a 'well-known string quartet', or a 'celebrated soprano', but should also intersperse popular artists—a 'jazz-band, a virtuoso on the saw, a café-concert singer, the Pleyela, or a Negro at a bar'.[59]

This enthusiasm for an unconventional approach to concert-programming reflects an irreverence that gravitated naturally towards the satire of popular spectacles, especially satires in which circus clowns poked fun at the self-serious professional musician. In *Études*, Milhaud describes a sketch called 'The Cello Parody' in which the Pompoff clowns constructed a makeshift cello out of pigskin, rope, and a broom, and proceeded to deliver a mock performance of Romantic melodies. The purpose, Milhaud observes, was to ridicule both the long-winded, effusive nature of nineteenth-century melodies and the performance style of ardent musicians who 'let their heart overflow on the first string of their Stradivarius'. Milhaud was also fond of a sketch by the Fratellinis in which they tossed music to the ground and donned enormous socks instead of batons in a parody of traditional performance preparations. The brilliance and humour of these clown acts, he explains, lay in their 'very agreeable caricature' and in their presentation of the 'dear Muse [of music] in a manner which we are not accustomed to encountering in concert halls'.[60]

Milhaud's distaste for the long melodies of nineteenth-century music and for the impassioned playing style which characterized performances of this repertoire was offset by a love for eighteenth-century classicism. In his writings, he often referred to the 'real French tradition'[61] of Rameau and to the 'simple clear art'[62] introduced by Mozart and Scarlatti. Milhaud's argument that contemporary French music should escape the burden of Wagnerian Romanticism by recovering this eighteenth-century tradition closely resembled the neo-classical views of composers in the 1920s. For Milhaud, however, classical ideals of simplicity, balance, proportion, and restraint were not simply associated with Bach, Rameau, and other composers of the eighteenth century. He imposed these ideals on the popular musics he loved. Thus American syncopated dance music, blues, and Parisian music-hall songs became models of

---

[59] *Études*, p. 69.

[60] For Milhaud's description of these satirical sketches, see *Études*, pp. 79–80.

[61] Quoted by John Alan Haughton in 'Darius Milhaud: A Missionary of the "Six"', *Musical America* 27 (Jan. 1923): 3.

[62] *Notes without Music*, p. 80.

classicism as well as inspirations for contemporary composers who wished to incorporate the eighteenth-century ideal.

By identifying his favourite French and American popular idioms with features of classicism, Milhaud granted popular music an exalted status. In his description of the blues singer whom he heard perform at the Capitol Club in Harlem, Milhaud remarked upon the poignancy of her melody and pronounced the melodic contour 'as pure as any beautiful classical recitative'. In his discussion of performances by Wiéner and Lowry at the Bar Gaya, he noted that the shift from ragtimes and foxtrots to works of Bach was not startling for the listener since both syncopated music and the music of Bach demanded rhythms of a strict 'inexorable' regularity. Even when Milhaud praised the restraint and discipline of popular idioms without specifically invoking Bach or the term 'classical', the link between simplicity and classicism was implicit. For instance, he praised Billy Arnold's band for the 'perfect tact, proportion, and balance' of its performance style. He admired the Parisian music-hall singer Damia because she performed with lightness and decorum and understood the value of a single, simple gesture.[63]

The classical traits of economy of means and proportion which Milhaud invokes in his descriptions of Billy Arnold and Damia also appear in his statements on Satie and the young French school. For example, in an article written in 1923 for the *North American Review*, Milhaud speaks of Satie's 'concision and soberness' and then praises young French composers like Poulenc and Auric who had returned to 'soberness of expression' and 'simplicity of harmonic line'.[64] Since Milhaud viewed American popular dance music and Parisian music-hall entertainment as the contemporary embodiment of classical ideals, he may have felt that Satie, Poulenc, and Auric learned their lessons in simplicity first and foremost from the popular arena. Yet his many-sided and ambiguous interpretation of simplicity also suggested a basis in French entertainment of the past. Milhaud intimated in his writings that the roots of Parisian fair and music-hall lay in the simplicity of French folklore. He associated this folklore with nostalgia and tenderness, just as Poulenc and Auric depicted the appealing sentimentality of Parisian street life, and he praised Satie for recovering French simplicity and 'nostalgia' in *Parade*:[65]

---

[63] Quotations on the Capitol Club singer, on Billy Arnold and on Damia taken from *Études*, pp. 59, 52, 86. Quotation on Wiéner and Lowry from *Notes without Music*, p. 103.

[64] Milhaud, 'The Evolution of Modern Music in Paris and in Vienna', *North American Review* 217 (Apr. 1923): 549.

[65] *Études*, p. 16.

After the blow of the *Rite*, Satie recognizes himself and brings music back to simplicity, thus opening the way for the young musicians who will form the post-war French school. He then gives us *Parade*, a ballet created in collaboration with Cocteau, Picasso, and Massine, one of the most beautiful successes of the Ballets-Russes, where the nostalgia of the music-hall, transposed, offers us a totally unsuspected art.

Satie's revival of simplicity and nostalgia inspired Poulenc and Auric to join him in creating a new French art music. In sketching this process, Milhaud brings together the associations of simplicity with classicism, popular entertainment, and French folklore:[66]

Following in the path of Satie, [Poulenc and Auric] rediscovered the folklore of France and particularly of Paris. The sadness of fairs, of distant bands, of the music-hall find in Auric an echo which is often bitter and incisive, sometimes brutal and full of rhythmic life. His score of *Les Fâcheux . . .* is among the most significant of Auric's works in its vividness, its candour, and its clarity of writing and of thought.

Statements by Milhaud, Poulenc, and Auric which defied Wagnerian 'sublimity' and championed instead a simple, down-to-earth music had their source in Cocteau's manifesto of 1918, *The Cock and the Harlequin*.[67] In this pamphlet Cocteau made musical simplicity a central theme, defining it in terms of aphorisms. Many of the aphorisms celebrated the everyday nature of simplicity and contained references to 'earth', 'walking', and 'building' which opposed the ethereal sounds of Impressionism. For example, in an attack on Impressionist music, Cocteau declared, 'Enough of clouds, waves, aquariums, watersprites, and nocturnal scents; what we need is a music of the earth, everyday music'. A similar denouncement of 'music one swims in' concluded with a demand for 'MUSIC ON WHICH ONE WALKS'. Moreover, Cocteau's dismissal of 'hammocks, garlands and gondolas' culminated in the statement, 'I want someone to build me music I can live in, like a house'.[68] In this last mandate Cocteau endorsed not only the everyday but the useful, practical aspect of musical simplicity. By usefulness, he seems to have meant the ability of music to fulfil a function in everyday life, a role expressed particularly well and concisely in the aphorism, 'MUSICAL BREAD is what we want'.

The third aspect of simplicity—the absence of 'superfluity' and

---

[66] *Études*, p. 18. Milhaud's reference to the 'sadness' of fairs parallels the references of Poulenc and Auric to a feeling of melancholy beneath the surface gaiety of popular institutions.

[67] For attacks on the 'sublime', see the discussion of Georges Auric in my Introduction.

[68] Unless otherwise specified, all quotations from *The Cock and the Harlequin* are taken from Rollo Myers's translation in *A Call to Order* (New York: Haskell House, 1974).

'frills'—could be translated into musical terms. In the following aphorism, for example, Cocteau indicated that by opposing superfluity he was embracing a music based on sparsity of texture and economy of means: 'A POET ALWAYS HAS TOO MANY WORDS IN HIS VOCABULARY, A PAINTER TOO MANY COLOURS ON HIS PALETTE, AND A MUSICIAN TOO MANY NOTES ON HIS KEYBOARD.' Lack of superfluity had other specifically musical connotations. In his defence of sparse textures, Cocteau emphasized the importance of melodic line: 'In music, line is melody. The return to design will necessarily involve a return to melody.' As an alternative to the 'long-drawn-out works' of Wagner, moreover, which deceived the audience with 'lies' and 'hypnotism', Cocteau proposed the 'short, simple, truthful work'.

Cocteau's statements defending musical simplicity were inspired by a particular model: the music of Erik Satie. Indeed Cocteau derived his definition of simplicity from the elements he heard in Satie's music. Although his aphorisms did not always refer to Satie, he wove statements on the composer around slogans championing a down-to-earth, concise, tuneful music, and thus the link was implicit. The principal aphorisms about 'music on which one walks' and 'music one lives in', for example, were preceded by the following: 'Satie teaches what, in our age, is the greatest audacity, simplicity. Has he not proved that he could refine better than anyone? But he clears, simplifies, and strips rhythm naked.'

In praising Satie's brevity and linearity, Cocteau drew a link between the restraint of an everyday musical style, the simplicity of Satie, and principles of eighteenth-century classicism. On several occasions in *The Cock and the Harlequin*, Cocteau employed the term 'classical' in his discussions of Satie. In one instance he defended 'Satie's classicism and his respect for the Schola', and in another he mocked Debussy's 'feminine grace' and praised Satie for continuing to 'follow his little classical path'. Such connections between simplicity, the music of Satie, and classicism established a precedent for a similar equation set forth in the writings of Milhaud and Poulenc. Moreover, the tendency of Milhaud, Poulenc, and Auric to invoke classical principles in their accounts of popular entertainment reflected the influence of Cocteau's manifesto. In the preface to *The Cock and the Harlequin*, Cocteau cited techniques of classicism which, he argued, could be gleaned from a knowledge of the circus and the music-hall:[69] 'What I was going to look for at the circus and the music-hall, it was not as people have so often claimed, the charm of clowns and negroes, but a lesson in equilibrium. A school of hard work, of discreet force, of useful grace, an elevated

[69] See Cocteau, 'Préface' to *Le Coq et l'arlequin*, in *Le Rappel à l'ordre*, p. 10.

school which separated me from inattentive minds.' The traits mentioned by Cocteau—'hard work', 'discreet force', 'useful grace'—suggested not only principles of classicism, but the down-to-earth, economical features of simplicity which he championed throughout his manifesto.

The association of popular spectacles with notions of usefulness and the everyday was made on several other occasions. Cocteau applauded the 'noise' and 'cataclysm' of 'jazz-bands', for instance, because such 'crudity' of sound served to 'wake people up'. The conviction that people must be roused from their torpor by the 'savagery' of popular entertainment was reaffirmed just a few years later when Georges Auric described how people were awoken by the 'uproar' of circus, fair, and music-hall. In both *The Cock and the Harlequin* and the *Carte blanche* articles, moreover, Cocteau portrayed popular milieux as distillations of life itself. His argument in *Le Coq* that 'music-hall, circus, and American negro bands . . . fertilize an artist just as life does', can be placed alongside a passage in *Carte blanche* in which he described how the movie screen recreated life by displaying quick successions of events, simultaneous occurrences of several different actions, and a teeming abundance of people and forms.[70]

An aspect of the popular realm addressed in *The Cock and the Harlequin* and in an article written by Cocteau for *Vanity Fair* in 1917, and one which also emerged in the writings of Milhaud, Poulenc, and Auric, was the feeling of nostalgia evoked by popular institutions. In a discussion of Satie's *Parade* in *The Cock and the Harlequin*, Cocteau explained that the fairground setting of the ballet, and the circus and music-hall characters who enacted their roles in this setting, represented 'varieties of nostalgia hitherto unknown'. Similarly in *Vanity Fair*, he described how the acrobatic act in *Parade* combined 'the accidental art of the circus with the happy remembrances of childhood'.[71] Implicit here was a belief that Parisian popular entertainment created a feeling of nostalgia because of its association with the past. Nostalgic fondness for the past, moreover, was clearly linked with Cocteau's theme of the everyday, since the milieux which produced nostalgic emotions were precisely those everyday spots embraced by Cocteau: 'circuses, music-halls, carousels, public balls, factories, seaports, and movies'.[72]

---

[70] Cocteau, *Carte blanche* in *Le Rappel à l'ordre*, pp. 92–3. In a *Carte blanche* article of 7 April 1919, Cocteau described circus clowns who 'give us in ten minutes, a condensation of life . . .' See *Carte blanche*, p. 83.

[71] Jean Cocteau, '*Parade*: Ballet Réaliste: In Which Four Modernist Artists Had a Hand', *Vanity Fair* 9 (Sept. 1917): 37.

[72] See Cocteau's letter to Stravinsky on 11 August 1916, reprinted in Robert Craft, ed., *Stravinsky: Selected Correspondence* 1 (New York: Alfred A. Knopf, 1982), p. 87.

Cocteau's comments on nostalgia set the stage for Milhaud's discussions of popular entertainment and the simplicity of French folklore, and for Poulenc's and Auric's fascination with popular milieux as emblems of the past. In addition, Cocteau's argument that by incorporating sounds of the popular realm Satie's *Parade* successfully married 'the racket of a cheap music-hall with the dreams of children and the poetry and murmur of the ocean',[73] suggests a source for the theme of seriousness beneath a surface gaiety, which preoccupied Poulenc, Auric, and to a lesser extent Milhaud. With the statements on *Parade*, then, we come full circle: a description in *The Cock and the Harlequin* of the 'brevity', the 'unromantic melancholy', and the 'touching simplicity' of Satie's ballet encapsulates the link between popular entertainment and the traits of classicism and simplicity which Cocteau was advocating.

When Poulenc, Auric, and Milhaud described the nostalgia of popular institutions, they were thinking specifically of Parisian milieux. Cocteau's discussions of nostalgia made no reference to Paris, yet in his account of Mistinguett which appeared in the *Portraits-souvenir* of 1935, he associated his favourite singer with the vitality of Parisian street cries and, more generally, with the 'best aspects of Paris'.[74] Such patriotism was also reflected in a statement in *The Cock and the Harlequin* which praised the cultural purity of Parisian café-concert entertainment: 'In the midst of the perturbations of French taste and exoticism, the café-concert remains intact in spite of Anglo-American influence.' Although it is difficult to reconcile this strong allegiance to Paris with the love which Cocteau expressed on other occasions for American syncopated music, in the opening issue of *Le Coq* (May 1920) Georges Auric explained why, in his view, an interest in jazz was especially important for French composers during the late 1910s. In an article which he entitled 'Bonjour Paris!', Auric argued that jazz had served a crucial function in offering French composers a fresh source of inspiration and an alternative to the 'seductions' of Wagner and Debussy. Auric went on to state that now that French composers had been shaken from their torpor, it was time to bid New York farewell and say hello to Paris.[75] He did not define the new French music, but the implication was a music based on French sources which preceded Debussyism.

Cocteau and Auric were clearly the principal proponents of this brand of nationalism, for such cryptic mottos as 'Return to poetry.

---

[73] Cocteau, '*Parade*: Ballet Réaliste', p. 106.

[74] Cocteau, *Portraits-souvenir*, p. 137.

[75] Auric, 'Bonjour Paris!', *Le Coq* 1 (May 1920). Significantly, with its third issue (July, August, September, 1920), *Le Coq* was renamed *Le Coq parisien*.

Disappearance of the skyscraper. Reappearance of the rose', which Cocteau wrote in praise of nationalism and against America, appeared in the same May 1920 issue of *Le Coq*. Francis Steegmuller, Cocteau's principal biographer, explains that the rose invoked by Cocteau was the rose of France. According to Steegmuller, the symbolism was twofold: first, a rose appeared as the emblem on François Bernouard's printing press, which published *Le Coq*; second, the rose was the central image of an ode by the sixteenth-century French poet Ronsard.[76] As a tribute both to Bernouard and to Ronsard, Cocteau wrote a pastiche of Ronsard's ode, which he called 'La Rose de François', and printed it in *Le Coq*. It was characteristic of Cocteau's love for puns that 'François' referred not only to the publisher Bernouard but to the French nation.

Darius Milhaud expressed a similar allegiance to France and even praised Satie for rediscovering the 'voice which belongs to French music freed of every foreign influence';[77] yet in 1920 he did not echo Auric's call for a renunciation of American popular music. Milhaud probably felt that the foreign influence which posed a threat was not American but German. He did, however, defend his interest in popular idioms against criticisms that he was a comic composer. In *Notes without Music* he described how annoyed he was after the première of *Le Bœuf sur le toit*, when the public placed his music in a 'music-hall circus system of aesthetics' and labelled him a 'showground musician'. Such a response instilled in Milhaud an ambivalence towards the popular realm and a need to uphold the serious nature of his art compositions. In the June 1920 issue of *Le Coq*, he declared in a brash motto, 'I want to write eighteen quartets.' The claim that he would write one more quartet than Beethoven was intended as a defence of serious chamber music in the face of the music-hall aesthetic. Yet it is difficult to assess how completely Milhaud wished to detach himself from this aesthetic in 1920, given the defiant humour of his motto and its resemblance to aphorisms by Cocteau and Satie.[78]

Francis Poulenc, by contrast, was unabashed by his light-hearted or 'street music side'. Poulenc's references to popular entertainment differed from Milhaud's and Auric's, moreover, in their exclusive focus on Parisian institutions. He did not find it necessary to bid

[76] For this interpretation of Cocteau's numerous references to the 'rose' in *Le Coq*, see Francis Steegmuller, *Cocteau*, pp. 259–60.

[77] *Études*, p. 20.

[78] For Milhaud's explanation of this motto, see Milhaud, *Entretiens avec Claude Rostand* (Paris: René Julliard, 1952), p. 90.

New York farewell in 1920 because, according to him, the source of popular influence on his music was consistently French.

The many statements offered by Poulenc, Auric, and Milhaud on the lure of popular entertainment acquire special significance when considered alongside the pieces which the three musicians composed during the years of the Saturday evening excursions and the gatherings at the Gaya and the Bœuf sur le toit. Diversity, simplicity, parody, the down-to-earth, and the bittersweet tone of nostalgia were principles which characterized not only the composers' aesthetic but their musical compositions. Moreover, the seriousness of Auric's dismissal of American jazz, Milhaud's denial of a comic style, and Poulenc's exclusive allegiance to French popular music can be tested by careful study of the musical compositions written in the seven-year span following Satie's *Parade*. Since *Parade* launched the popular language and the new aesthetic, an analytic discussion best opens with Satie's ballet.

5

Ragtime, Diversity and Nostalgia: The Language of Satie's Parade

THE impact of *Parade* was felt immediately. By 1918, with the publication of Cocteau's pamphlet *The Cock and the Harlequin*, Satie had assumed his position as the father of a new French movement that opposed Wagner and championed simplicity and everyday themes. Many of the ballet's elements of varied repetition, musical quotation, brevity, and spoof were familiar from Satie's piano pieces of the 1910s. Yet *Parade* was Satie's first concert work to explore a popular language and his first to become known to a large French public. The close attention lavished on the ballet by young artists, poets, and composers in Paris suggests the need for a musical analysis, one that will reveal Satie's experimentation with new techniques of diversity and nostalgia.[1] This analysis will also strengthen our understanding of *Parade* as an inspiration and a model for the songs, instrumental pieces, and ballets that Milhaud, Poulenc, and Auric composed in the following years.

The ballet's scenario, though simple in its story-line, is ambitious and unconventional in its medium and resembles a libretto more than a synopsis. Consisting of detailed notes and telegraph-style descriptions designed to introduce the different music-hall characters, the 'scenario' was delivered to Satie by Cocteau in May 1916. As he composed his score, Satie worked closely with Cocteau's material, responding to the individual personality of each character, the nostalgic component of the story and the medley of popular venues. Cocteau based the scenario on a particular category of *théâtre forain*, the *théâtre de magie-variété-music-hall*, which combined music-hall numbers (clowns, acrobats, magicians, and animal-tamers) with dramas and *féeries* and placed these spectacles in a fairground setting.[2] The action of the ballet occurs on a *parade* featuring a French and an American Manager who aggressively

[1] The musical language of *Parade* has never been analysed in a systematic way that accounts for the different melodic types that Satie chose and for the music's popular sound. No doubt, *Parade* is a very difficult work to analyse, since it is not unified in a traditional sense.

[2] Richard Axsom deduces that the category of *théâtre forain* selected by Cocteau was the *théâtre de magie-variété-music-hall*. See *'Parade': Cubism as Theater* (New York: Garland

shout ballyhoos to introduce three music-hall numbers: the Chinese Magician, the Acrobats, and the Little American Girl.[3] [See Plates 6–8.] The three acts are vehicles which the Managers hope will amuse the crowd and entice them inside the theatre to see the actual show. Cocteau fashioned the Chinese Magician after the popular conjurors whose decapitation, transformation, and ventriloquy acts were the principal attraction at the *théâtre de magie-variété-music-hall*. The Little American Girl was inspired by a *théâtre forain* act entitled 'Bib and Bob, the very funny American pantomime', which combined dancing and pantomime.[4] The Acrobats were staples of the *théâtre de magie-variété-music-hall*.

Of the three music-hall acts, the Little American Girl was the most complex: Cocteau created her to symbolize American life as it was then known to the French. He gleaned his knowledge of American popular culture and technological inventions primarily through contact with the American films shown in Paris. Thus the character description he sent to Satie assembled images drawn from the films he had seen. The impressive variety—film stars and steamships, posters, skyscrapers, a gospel hymn, and the Brooklyn Bridge—and the excitement generated by their steady accumulation, conveys an enchantment with the United States that was typical of French sentiment during and after World War I:[5]

The *Titanic*—'Nearer My God to Thee'—elevators . . . steamship apparatus—*The New York Herald*—dynamos—airplanes . . . palatial cinemas—the sheriff's daughter—Walt Whitman . . . cowboys with leather and goat-skin chaps—the telegraph operator from Los Angeles who marries the detective in the end . . . gramophones . . . the Brooklyn Bridge—huge automobiles of enamel and nickel . . . Nick Carter . . . the Carolinas—my room on the seventeenth floor . . . posters . . . Charlie Chaplin.

Cocteau also wrote detailed stage directions for the 'Steamship Ragtime' of the Little American Girl, and these recall the action of a Mary Pickford or Pearl White film in their dizzying display of a

Publishing, 1979), p. 36. Axsom gleans his information on the Parisian fairground entertainment that inspired Cocteau's scenario from Jacques Garnier, *Forains d'hier et d'aujourd'hui* (Orléans: Les Presses, 1968), pp. 274–87.

[3] The third Manager was the two-man horse which appeared in between the Little American Girl's music and the Acrobats'. Cocteau called for numbered placards announcing each act, in imitation of the placards preceding acts at the music-hall.

[4] For information on the *théâtre forain* acts featuring 'Bib and Bob' and the magicians, see Garnier, *Forains d'hier et d'aujourd'hui*, pp. 284, 286.

[5] Cocteau wrote character descriptions for all three music-hall numbers. The portrait of the Little American Girl, from which the following excerpt is taken, is published in translation by Frederick Brown in *An Impersonation of Angels* (New York: The Viking Press, 1968), pp. 128–9.

great range of activities: riding a horse, catching a train, driving a Model T Ford, swimming, and playing cowboys and Indians.[6]

The light-heartedness of the 'Steamship Ragtime', of the images comprising the portrait of the Little American Girl, and of the ballet's fairground setting is counterbalanced by the sadness of the story.[7] In Cocteau's scenario the *parade* performers are unsuccessful in their attempts to promote the show that will take place inside the theatre, because the spectators mistake the *parade* for the real performance. The Managers become increasingly desperate and 'vulgar' as they ballyhoo and urge the audience to come inside, but their efforts are ignored. No one enters the theatre, and when the last music-hall number is over the Managers collapse in exhaustion. The Chinese Magician, the Acrobats, and the Little American Girl then reappear and try for their part to explain that the show will take place inside. The disparity between the lively, energetic stunts of all three music-hall numbers and the unwillingness of the audience to enter the theatre is poignant. Cocteau's theme of the unsuccessful sideshow reveals his vision of a nostalgic, melancholy core beneath the surface charm of popular entertainment.

The poet Guillaume Apollinaire responded to *Parade* by promoting what he felt to be the ballet's extraordinary novelty. On 11 May 1917 he predicted in the newspaper *Excelsior*, and then one week later in the ballet programme, that *Parade* would be the first in a succession of manifestations of the 'esprit nouveau' or new spirit. The important word here was 'new'. Apollinaire believed that, unlike ballets before it, *Parade* created a new and radical union between painting (designed by Picasso) and dance (Leonide Massine's contribution). *Parade* also introduced a 'realism' which Apollinaire equated with the Cubism of Picasso. Realism did not mean an exact rendering of truth and of outward appearances, but rather a suggestion, a synthesis, a reconciliation of contradictory elements which brought with it the experience of surprise. Apollinaire's note concluded with a tribute to the contradictions and to the fusing of commonplace and fantastic elements in the musical-stage piece that Cocteau termed his 'ballet réaliste': '*Parade* will upset the ideas of quite a number of spectators . . . A magnificent music-hall Chinaman will give free rein to the flights of their imagination, and the American girl, by turning the crank of an imaginary automobile,

---

[6] Richard Axsom learned during an interview with Leonide Massine that Cocteau's Little American Girl was based on Mary Pickford, a favourite American star during the war years. See Axsom, *'Parade': Cubism as Theater*, p. 42.

[7] For Cocteau's synopsis of the story, see the preface to the four-hand piano version of *Parade*.

will express the magic of everyday life, while the acrobat in blue and white tights celebrates its silent rites with exquisite and amazing agility.'[8]

SIMPLE TUNEFULNESS

The simple, spare musical language of *Parade* parallels the simplicity of the scenario and reinforces Satie's songful melodic style, with its evocation of music-hall and fairground. Satie accompanies the stunts of the Managers and the three music-hall entertainers with a multiplicity of short tuneful ideas often placed above a note-chord accompaniment and repeated so as to become memorable. Melodic types range from two-, three-, and four-note ostinati to longer melodic gestures comprising distinctive entities but usually lacking a strong sense of melodic and tonal closure. With this array, Satie offers his personal re-creation of features of late nineteenth- and early twentieth-century French and American popular music. The brevity of the successive melodic fragments enhances their contrast and mirrors the diversity of acts and the rapidity with which one act succeeded another at the Parisian music-hall, café-concert, circus, cabaret, and fair.

Ostinati and pendulum figures

Among the various figures contributing to the musical fabric of *Parade*, the ostinato and the pattern in which two pitches rotate back and forth—henceforth referred to as the pendulum—are the most pervasive. Satie uses simple ostinati to introduce the principal themes of sections such as the Chinese Magician or the Acrobats. He heightens the tuneful quality of these figures by placing them in the highest register of a stratified texture in which several lower-voice ostinati act as tonal supports. Frequently Satie presents his ostinati in chains of four-bar phrases so that one repeated pattern is followed immediately by another.

An early passage of the Chinese Magician illustrates Satie's treatment of the tuneful ostinato (see Ex. 5.1). The passage opens with the English horn's introduction of a figure (bars 1–4) containing four even quarter notes that ascend a whole step and a minor third and then descend a minor third to form an arch-like contour. The English horn figure occupies two bars, sounds in conjunction with its embellished version in the violas, cellos, and

---

[8] For a translation of Apollinaire's complete essay 'Parade', see Appendix VII in Francis Steegmuller, *Cocteau: A Biography* (Boston: Little, Brown, & Co., 1970), pp. 513–14.

Ex. 5.1. Ostinati in *Parade*

Man Ray, *Erik Satie*, March 1924, photograph Man Ray Trust, Paris.

2. Louis Marcoussis, *Darius Milhaud*, 1936, engraving Musée national d'art moderne, Paris.

Man Ray, *Francis Poulenc*, 1922, photograph Man Ray Trust, Paris.

4. Man Ray, *Jean Cocteau*, *c.*1924, photograph Man Ray Trust, Paris.

5. Pablo Picasso, overture curtain for *Parade*, 1917, tempera on canvas, 10.6×17.25 m., Musée national d'art moderne, Paris.

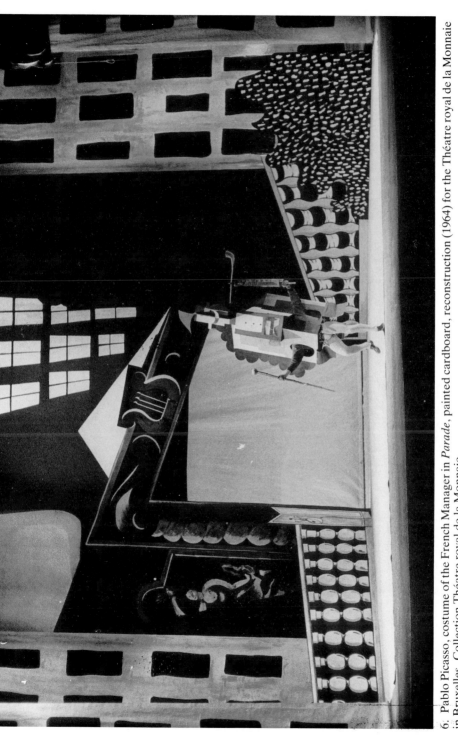

6. Pablo Picasso, costume of the French Manager in *Parade*, painted cardboard, reconstruction (1964) for the Théatre royal de la Monnaie in Bruxelles, Collection Théatre royal de la Monnaie.

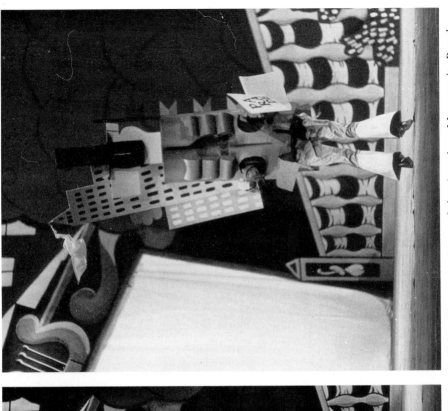

8. Pablo Picasso, costume of the American Manager in *Parade*, 1917, painted cardboard, reconstruction (1964) for the Théatre royal

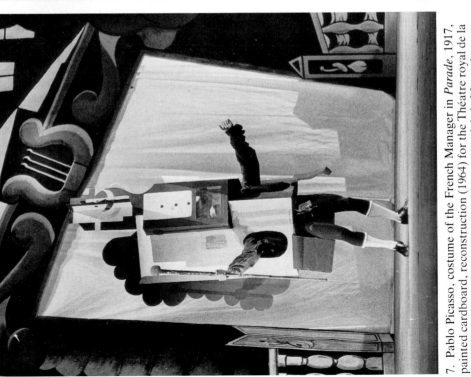

7. Pablo Picasso, costume of the French Manager in *Parade*, 1917, painted cardboard, reconstruction (1964) for the Théatre royal de la

Jardin de Paris, a poster by Jules Chéret, Musée de la Publicité, Paris.

10. Fernand Léger, stage model of *La Création du monde*, 1923, The Dance Museum, Stockholm.

11. Fernand Léger, king's costume design for *La Création du monde*, 1923, water-colour, 44×24 cm., The Dance Museum, Stockholm.

12. Henri Toulouse-Lautrec, *The Trapeze Dancer at the Cirque Médrano*, 1888, charcoal on paper, 37×28 cm., Musée Toulouse-Lautrec, Albi.

harp, and is repeated once so that a four-bar phrase is established. Then in the following bar (bar 5) the clarinet introduces a new ostinato. This idea likewise contains four pitches, but is only one bar long and acquires its tuneful quality through an angular melodic contour coupled with persistent dotted rhythms. Satie repeats the ostinato twice and follows it with a bar of sustained sound, thereby completing a second four-bar phrase. Then he moves on to his third figure, in the E♭ clarinet (bar 9). This pattern, the simplest of the three, is a pendulum in which C♯ and D♯ rotate in the clarinet, and the same pitches—ornamented by an F♯ pedal tone—are heard in the violins. The tunefulness of the pendulum is heightened by the familiarity of its sound: it represents a transposition of the first two pitches of the English horn ostinato (Ex. 5.1, bar 1). After repeating the figure three times and establishing a four-bar phrase, Satie introduces a new theme in the horns. The pendulum continues to be heard but its role shifts from melody to accompaniment.

In this way, from bars 1 to 12 Satie presents a series of three melodic gestures that lead to a horn theme in the Lydian mode on E (bar 13). In retrospect all three can be interpreted as a preparation for this horn theme, and the first and the third actually lie within the Lydian mode. At the same time, the three figures are independent ideas; Satie defines their melodic role by placing them within a layer of ostinati and then creating a hierarchy in which each of the three in turn predominates. For instance in bars 1–4, where the English horn presents its four-note idea, the bassoons play a pedal point of repeated Bs that constitutes a tonal support, and the double bass decorates the bassoon line and forms a pendulum by alternating B with G in a four-note pattern (B–G–B–G). Similarly the pendulum idea that begins in bar 9 in the E♭ clarinet is supported by repeated linear octaves in the cellos. Satie's choice of four-bar phrasing contributes to the tunefulness of the three ideas, since the four-bar divisions create a symmetrical layout that allows the listener to sense how many times an idea will be repeated and when a new pattern will be introduced. The resemblance of the supportive lower-voice ostinati to the ostinati in *The Rite of Spring* suggests that popular models were not Satie's only source of influence.

The ostinati that embellish the C♯–D♯ pendulum and provide tonal support, as well as the C♯–D♯ idea itself, furnish a backdrop for Satie's presentation of the Lydian theme in the horns. In between thematic statements, Satie introduces new ostinati that assume important melodic roles, each time for a period of four bars and each time supported by lower-voice ostinati (linear octaves or two alternating pitches). In Ex. 5.2*a*, for instance, the English horn plays a simple six-note idea consisting of two whole-step ascents first from the pitch A to C♯, then from B to D♯. Ex. 5.2*b* illustrates an ostinato comprised of a *Klangfarben* line (melody of instrumental tone colours) in which the English horn begins with a linear octave on B, and the B♭ clarinet follows with a linear octave on C♯. The ostinati in both examples sound tuneful not only because of their scalar basis but because the rotation of two pitches a step apart is familiar to the listener from the pendulum first introduced in the E♭ clarinet (see Ex. 5.1, bar 9).

Ex. 5.2. Tuneful ostinati in *Parade*

(a)

(b)

The ostinato that Satie uses to prepare the Acrobats' theme differs in several ways from the series of three figures that introduce the horn theme in the Chinese Magician's section. Scored for xylophone and strings, the Acrobats' ostinato (Ex. 5.3, bar 1) is tuneful not only because it falls into four-bar phrases, but because Satie expands its length so that the arch-like contour is quite audible. He selects an intervallic succession of ascending whole-step followed by two perfect fourths that ascend and then descend (F♯–G♯–C♯–F♯–C♯–G♯) and embellishes this outline by repeating each pitch at the lower octave before moving on to the next. In this way the figure occupies four bars and is heard twice. The ostinato's length enables Satie to create a musical genre, in this case a waltz. Within the triple metre setting, accents fall on the first quarter-note of each bar, and the ostinato is heard above a note-chord accompaniment in which the tonic pitch C and the dominant pitch G alternate with two repeated tonic chords. Satie derives this accompaniment from the characteristic note-chord patterns used in French music-hall waltz songs and more generally in instrumental waltzes. His placement of tonic and dominant pitches on downbeats of the bar and his emphasis on the two as tonal poles are equally characteristic.

The Managers' passage of *Parade* presents another accompaniment derived from French and American popular songs. This time

Ex. 5.3. The Acrobats' ostinato in *Parade*, Satie's four-hand piano reduction

the accompaniment is a three-note linear pattern in which tonic and dominant pitches alternate (see Ex. 5.4, bars 1–4). The pattern closely resembles a four-note tonic-dominant alternation which appears in the accompaniment of the French music-hall song 'Valentine' (see Ex. 5.5). Satie's figure differs, however, in its triple rather than duple metre. In addition Satie departs from conventional popular patterns by bringing his figure back identically as an

Ex. 5.4. The Managers' ostinato in *Parade*

Ex. 5.5. The ostinato-like accompaniment in "Valentine"

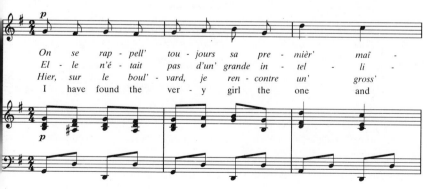

ostinato rather than employing new pitches in accordance with the
tonal scheme of the melody.[9] The use of the ostinato here and
throughout *Parade* illustrates Satie's subtle means of distancing
himself from and transforming the popular idioms he evokes.
Stravinsky's inventive treatment of the ostinato in *Petrushka* (1911),
The Rite of Spring (1913), *Three Japanese Lyrics* (1913), *Three
Pieces for String Quartet* (1914), *Pribaoutki* (1914) and other works
of the early 1910s furnished Satie with an important model.

Initially the three-note linear ostinato in the Managers' section is
heard four times alone in a monophonic texture with piccolo
duplicated by harps, violins, and double bass (Ex. 5.4, bars 1–4).
Then in the fifth bar the Managers' theme begins in the clarinet,
piccolo drops out, and the ostinato persists in the remaining
instruments as an accompaniment. The imitation of popular music is
evident not only in the tonic-dominant alternation of the ostinato
(E–B) and the four-bar phrasing, but in the resemblance of the four-
bar introduction (bars 1–4) to vamps in popular songs.[10]

Throughout *Parade*, Satie often enhances the repetitive quality
intrinsic to ostinati by using the device of sequence to build his
ostinato figure. In the ostinato that introduces the first theme of the
Little American Girl, for instance (see Ex. 5.6, bars 1–8), the
repeated whole and half step alternations in the first bar are treated

Ex. 5.6: The Little American Girl's ostinato and theme in *Parade*,
Satie's four-hand piano reduction

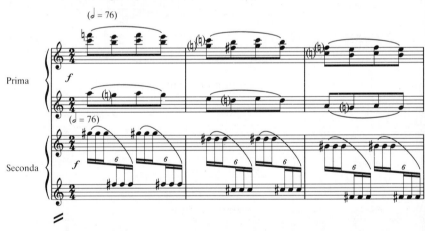

[9] For another example of Satie's use of a linear ostinato in the accompaniment, see bars
226–9 of *Parade*.

[10] For examples of two Tin Pan Alley songs that are introduced by a vamp, see Irving
Berlin's *That Mysterious Rag* and B. C. Hilliam's *Après la guerre*.

Ragtime, Diversity, and Nostalgia

sequentially and transposed down a perfect fourth in the second bar. Satie then repeats the two-bar ostinato at different octave levels before presenting the first theme of the Little American Girl. His use of sequence to construct ostinati, and the similarity between the intervallic content of these ostinati and the many sequential figures in *Parade* creates a repetitive sound typical of Satie's style and defiant of Romanticism.

Thematic figures

The melodic ostinati in *Parade* function most frequently as introductions to a longer fragment or principal theme identifying the Managers, the Chinese Magician, the Little American Girl, and the Acrobats. These themes represent a third melodic type that bridges the gap between ostinati and sequential gestures on the one hand, and melodies with distinctive contours and a sense of closure on the other. They resemble the ostinati in their restriction to a small number of different pitches (sometimes five or six), in their extension of material through repetition and sequence, and in their use of scalar and triadic shapes. They diverge from the repeated two- and three-note units and approach melody, however, in their use of variation techniques—as well as repetition or sequence—to extend material, in their occasional division into parts or phrases, and in their creation of a melodic profile with a distinctive beginning and end. With each theme, moreover, Satie creates a musical portrait of the particular music-hall act.

The Managers' theme is based on a simple three-note figure in the A clarinet (G–C–B) which Satie gives an arch-like contour so that it resembles the three-note ostinato heard in the accompaniment (see Ex. 5.7). Satie extends the Managers' figure and creates a theme by varying it rather than repeating it identically or applying sequence each time it returns. While the final two pitches remain constant, the first pitch is altered five times and in a few instances a new pitch is added at the end (bars 4, 6, 9). In this way, five different three- and four-note variants follow the opening figure of the Managers' theme. Satie does not establish a fixed order for these variants in which they succeed one another in the progression 1–2–3–4–5, and then recur in that order. Instead variant 1 returns a second time before all five variants have been presented, establishing the succession 1–2–3–1–4–5. After the five have been heard, moreover, Satie recalls four of them in the order 2–1–4–3–1–2, rather than returning to the original three-note figure and then repeating the variants in succession. Thus exact repetition is avoided both on a local level with the three-note figure, and on a long-term level with the five variants.

In the ostinati of *Parade*, melodic movement is repetitive and constant throughout, and beginning and end cannot be distinguished. In the Managers' theme, by contrast, melodic movement shifts at the end. The variant-figures cease and Satie brings back the opening three-note idea (see bars 10–11). His use of this figure to frame the successive variants creates a symmetry and suggests an ABA form. He establishes tonal closure, moreover, by repeating the figure twice and inserting an E before the opening pitch G (bar 10), so that the theme concludes unambiguously in E minor.

The varied repetition employed in the Managers' theme is by no means new with *Parade*. In piano pieces of the early 1910s such as *Peccadilles importunes* (1913), Satie extended material through the variation of melodic gestures which were often only four notes long. During the same years, in his *Three Japanese Lyrics* (1912–13), *Souvenirs de mon enfance* (1913), and *Three Pieces for String Quartet* (1914), Stravinsky worked with short melodic figures which

Ex. 5.7. The Managers' theme in *Parade*

he varied by reordering pitches, adding one new pitch, or altering the rhythm. In both cases, and again in *Parade*, the concision and simplicity of the melodic material create a repetitive quality suggestive of folk music. The appeal of folk song for both composers lay in its potential for forging a national voice as well as a voice of direct simplicity. For the Managers' theme, moreover, Satie cleverly applies subtle variation to a simple three-note figure and recalls the figures several times in order to mirror the Managers' repeated and varied attempts to entice the crowd inside. Since the changes in the first and last pitches do not affect the contour of the Managers' figure, the varied repetition acquires a circularity which Satie reinforces by recalling the opening figure and stating it three times at the end. The music's turning in on itself symbolizes the futility of the Managers' efforts.

For the theme of the Chinese Magician, Satie shifts from a diatonic to a pentatonic language which, among French composers, had become a characteristic sonority used to evoke the Far East. The discrepancy between the booming line in the trumpets and trombones which Satie uses to begin his pentatonic theme for the Chinese Magician (see Ex. 5.8, bar 1), and the delicate, exotic

Ex. 5.8. The Chinese Magician's theme in *Parade*

sounds of pentatonicism in Debussy's 'Pagodes' from *Estampes* (1903) or Ravel's 'Asie' from *Shéhérazade* (1903) is intended to mock the lure of the East. Satie deliberately substitutes booming brass for the soft piano and woodwind timbres of Debussy and Ravel, in order to parody the Impressionist captivation with 'Eastern' re-creations.[11]

The theme of the Chinese Magician, like the Managers', is a brief idea that offers certain similarities to the ostinati in *Parade* (Ex. 5.8, bars 1–10). It contains only five different pitches (B, D, E in the trumpets and trombone, bar 1, and G, A in the violins and B♭ clarinet, bar 2), its movement is triadic and scalar, and it features three figures that are repeated identically in the manner of ostinati during the course of the thematic statement (brass figure in bar 1, clarinet and violin figure in bars 2–3, clarinet and violin figure in bars 5–6). The theme resembles melody, however, in the antecedent-consequent relationship of its figures. The opening three-note brass idea ascends and is answered by the descending scalar figure in the clarinets and violins that completes the anhemitonic pentatonic scale centred on G. Dialogue is emphasized by Satie's placement of the two figures in different registers, his distribution of the two among different instruments, and his shifting of the rhythmic pattern from two eighth-notes followed by a quarter-note in the first figure to even eighths in the second.

Antecedent-consequent relations continue to operate in the following bars where the three-note brass figure serves again as antecedent (bar 4) and a second clarinet and violin figure, which is a variant of the first, provides the response. This variant marks the climax of Satie's theme. It occupies the midpoint, contains the highest pitches of the theme (D, E), and is followed by a return to the opening three-note figure. Satie establishes closure and distinguishes the end of the theme from the opening bars by repeating the three-note figure emphatically three times (bars 7–10). Tonal centring does not apply as it did in the Managers' theme, however, because although the ending centres around E, during the course of the theme Satie gives as much emphasis to the principal pitch of the pentatonic scale G as he does to the theme's final pitch E. The pitch G is reinforced, moreover, by the G–D alternation that comprises the accompaniment and appears as a cello ostinato throughout this section.

The Chinese Magicians' theme appears above a simple note-chord accompaniment in which the tonic pitch in the bass alternates

[11] Satie's pentatonic collection comprises G A B D E and is completed by the clarinets and violins.

with the dominant. The combination of the popular accompaniment, the steady duple time, and the antecedent-consequent relationship of the figures indicates that Satie has composed a march.

In the Acrobats' theme of *Parade* Satie also uses a dance model, this time a waltz (see Ex. 5.9, bars 1–12). The choice suits the requirements of Massine's choreography which called for more classical dancing and less pantomime than in any other section of the ballet. Of Satie's three themes for the music-hall acts, the Acrobats' bears the closest resemblance to melody. Its scalar movement echoes the ostinati of *Parade*, but its twelve-bar length and its use of three four-bar phrases differ strikingly from them. Satie introduces the theme with one of the ballet's longest ostinati, described earlier

Ex. 5.9. The Acrobats' theme in *Parade*, Satie's four-hand piano reduction

as a waltz tune. The theme's opening phrase (Ex. 5.9, bars 1–4, circled pitches) comprises a whole-tone descent in trumpets, trombones, violins, and cellos that begins on F and spans an octave. The second phrase (bars 5–8) departs from scalar descent in its third bar and introduces a new rhythmic pattern. In the final phrase (bars 9–12) Satie presents a varied return of the opening scalar descent, but stops on the third pitch of the scale and sustains it through the end of the phrase. Although the theme lacks tonal centring, Satie establishes closure by adhering to an ABA' formal pattern in the melody and by holding the pitch D (bars 11–12) rather than introducing rhythmic acceleration in the last two bars.

Satie's paraphrase of That Mysterious Rag

Ostinati, sequential figures, and longer melodic figures constitute the prevailing voice in *Parade*. The paraphrase of Irving Berlin's *That Mysterious Rag* which appears as a centrepiece in the Little American Girl's section represents a departure because it marks the only instance in *Parade* of a device employed frequently in Satie's earlier piano music: the paraphrase of a pre-existent source.[12] Berlin's song appears in a section of the Little American Girl's music entitled 'Ragtime du paquebot' or 'Steamship Ragtime'. This passage accompanies a series of actions inspired by the films of Mary Pickford and Pearl White; accordingly, the music is meant to sound quintessentially American.

Given the enormous popularity of cakewalks and rags in Paris since the time of Sousa's first appearance in 1900, an instrumental cakewalk might have seemed the likeliest choice. Yet Satie chose to base the 'Steamship Ragtime' on a Tin Pan Alley song which contained a few bars of simple syncopation, but lacked the untied and tied syncopation figures most characteristic of ragtime. If this choice seems surprising, it can be interpreted as a response to changing fashions in American popular music. *That Mysterious Rag* belongs to a category within the Tin Pan Alley style known as 'ragtime song' which first appeared in the early 1900s when songwriters began taking advantage of the ragtime craze and incorporating syncopated figures into their music.[13] Irving Berlin produced dozens of successful ragtime songs in the early 1910s, and

---

[12] For earlier examples of Satie's use of pre-existent sources, typically folk and children's songs, see the quotation of 'Le Roi Dagobert' in *Vieux séquins et vieilles cuirasses* (1913), and the use of 'Nous n'irons plus au bois' in *Regrets des enfermés* from *Chapitres tournés en tous sens* (1913).

[13] For information on ragtime songs, see Charles Hamm, *Yesterdays: Popular Song in America* (New York: W. W. Norton, 1979), pp. 317–21. A discussion of Berlin's ragtime songs appears on pp. 330–3.

the impact of this fashion was felt in France. Thus in 1917 when Satie chose a ragtime song as a musical model for *Parade*, he demonstrated his awareness that new musical styles had superseded the instrumental rag in popularity.

Satie's reworking of the Tin Pan Alley tune ranges from obvious borrowings of tonality, form, and rhythm to more subtle and sporadic duplications of pitch and melodic contour. The close knowledge of Berlin's song that enables Satie to paraphrase so many different musical parameters suggests that he may have possessed a copy of the sheet music. It is equally possible that he transcribed the song from a live performance.[14] In either case, Satie's adherence to the original is offset by compositional choices which distinguish 'Steamship Ragtime' from *That Mysterious Rag* and reveal his individual artistic stamp.

The C major tonality and the formal design of the Tin Pan Alley tune are transferred directly to the 'Steamship Ragtime'; this transferral may well have constituted Satie's first step in the paraphrasing procedure (see Exs. 5.10–11).[15] Ex. 5.10 illustrates that the 'Steamship Ragtime' is written in C major and divided into three parts. Satie preserves the length of each of Berlin's parts intact but reverses the succession of introduction-verse-chorus. Accordingly, the 'Ragtime' opens in C major with twenty-four bars corresponding to Berlin's chorus, moves to a sixteen-bar section based on the modulatory verse and closing on the dominant, and concludes in C major with an eight-bar section that paraphrases Berlin's introduction. Within each section, Satie's duplication of the rhythms of Berlin's melody is strikingly consistent. In addition, his bass-line rhythm coincides with Berlin's throughout the verse and occasionally in the chorus and introductory music as well (see bars 6, 7, 15, 16, 17, 22, 23 of chorus). Satie also adopts the eighth-note pairs that embellish the accompaniment in the chorus of *That Mysterious Rag*. These eighth-notes appear in all but one of the analogous measures of 'Steamship Ragtime' (see bars 1, 3, 4, 11, 12, 17, 19, 20 of Satie's chorus).[16] Thus Satie's rhythms, formal and tonal structure remain remarkably close to the Berlin model.

[14] Inquiries directed to SACEM in Paris have yielded no information on the appearance of a French edition of *That Mysterious Rag*. If Satie did learn the song from a sheet music copy, he must have obtained an American edition possibly brought to Paris during the war.

[15] In the analysis that follows, bar numbers refer to Exs. 5.10 and 5.11. Since the number of bars in verse and chorus of 'Steamship Ragtime' and *That Mysterious Rag* coincide exactly, but Satie's chorus precedes his verse, bars of verse and chorus are numbered separately in order to facilitate comparison. Thus bar 6 of Berlin's verse corresponds to bar 6 of Satie's. Introductions are not given bar numbers since in each case they duplicate the final eight bars of the chorus.

[16] The only bar in which Satie does not incorporate Berlin's syncopated eighth notes is bar 9.

Ex. 5.10. The "Steamship Ragtime" in *Parade*, Satie's four-hand piano reduction

The paraphrase of pitch material is more independent and takes three forms: first, duplication of pitches in the bass line; second, duplication of contours in the bass line; third, and more frequently, variation of melody and bass-line contours. Duplication of pitches occurs in only a few isolated measures in the bass line of the chorus. In the first three bars of the 'Steamship Ragtime', for instance, Satie models the descending octave movement from pitches C to B to A on the first two bars of Berlin's *Rag*. The turning figure in the second bar of 'Steamship Ragtime' ornaments the B octave and delays the descent. In bars 5–7, moreover, Satie's chromatic descent from E to D♭ duplicates the chromatic bass line descent from E to C♯ in bars 6–7 of Berlin's *Rag*.

The second form of pitch-paraphrase, duplication of bass-line contours, characterizes the verse of 'Steamship Ragtime'. In the first two bars, for example, Satie's movement up and then down by step to form a small arch (F♯–G–A♭–G) imitates the bass line's ascent and descent in the corresponding bars of the Tin Pan Alley verse (G♯–A–B–A). Satie also duplicates the contours of *That Mysterious Rag* in the third bar of his verse where the pitches on beats one and three of the bass line lie a minor third apart, in imitation of Berlin's lower voice, and the chords on beats two and four of the accompaniment are identical, as in Berlin's accompaniment.

Ex. 5.11. Irving Berlin, "That Mysterious Rag"

sneak-y freak-y   ev-er me-lo - di-ous   mys-ter-i-ous   rag ___   rag ___

Later bars of the verse of 'Steamship Ragtime', and indeed the entire chorus, offer variations rather than exact imitations of melody- and lower-voice-contours. The variation procedure in the verse includes both inversions and reorderings of the intervals used in the Berlin model. A comparison between the bass lines in the fourth bars of the verses of 'Steamship Ragtime' and *That Mysterious Rag*, for example, reveals that Satie has inverted the minor third so that it descends rather than ascends, and has placed it at the beginning of the bar rather than at the end. He retains the parallel ascending octaves. In the ninth bar of his verse, Satie derives his melody from an inversion of the melodic movement that appears in the corresponding bar of Berlin's verse. Throughout the chorus, moreover, Satie either inverts melodic contours (see bar 8) or varies them by imitating the melodic shapes while avoiding exact duplication of intervals (see bars 10–20, 22).[17] In both verse and chorus, he follows Berlin's principle of restating the same melodic idea in successive bars (see chorus, bars 13–14; verse, bars 1–2, 5–6). This repetitive device creates a dramatic aural link between Satie's rag and Berlin's original, even though the pitches themselves are different.

Indeed, while tonality, formal design, rhythm, and contour of melody and bass are transferred with slight modification from *That Mysterious Rag* to the 'Steamship Ragtime', Satie maintains a personal voice by introducing an entirely new set of pitches in the melodies of his verse and chorus. The bass-line pitches of 'Steamship Ragtime' coincide with those of Berlin's *Rag* only in the very few bars of the chorus discussed above. Such divergence in pitch material goes hand in hand with a choice of widely differing harmonies. In the course of the 'Steamship Ragtime', Satie's harmonies duplicate Berlin's only at three structural points: first, in

---

[17] The only exception is bars 3–4 of 'Steamship Ragtime' where Satie's melodic contour is identical to Berlin's.

bars 1 and 24 of the chorus where C major is used to frame the section; second, in bars 15–16 of the chorus where the dominant harmony leads the music back to the opening theme; third, in bar 16 of the verse where the dominant chord appears in order to signal the return of the chorus music. Even when harmonies coincide, moreover, as in bar 16 of the verse, Satie weakens the tonal orientation by delaying the arrival of V until the final beat of the bar and then placing the third of the chord in the bass.

A comparison of phrasing reveals Satie's delight in altering a simple popular tune by applying irregular bar groupings. Such asymmetrical phrasing is a consistent stylistic trait that can be traced to Satie's cabaret song 'Un Diner à l'Élysée' (1903), to *Trois Morceaux en forme de poire* (1890–1903) and onward to *Sports et divertissements* (1914). Interestingly, Satie chooses the chorus—the most regular, tuneful section of *That Mysterious Rag* and the section which opens Satie's 'Ragtime'—as the place to experiment. His verse, by contrast, adheres verbatim to the four-bar phrasing of his Tin Pan Alley model.

The phrasing of Berlin's chorus is highly symmetrical both in the length of bar groups and in the articulation of the phrase (see Ex. 5.12). It is based on a division into six four-bar units and a design in which the melody of phrases one, three, and five opens with a sustained whole note and closes with a series of two whole notes. Satie alters this design by extending his second phrase so that it does not conclude until bar 10. In this way, the whole note in bar 9 becomes the climax of phrase two, rather than the opening pitch of phrase three. Satie's third phrase then begins in bar 11 with the series of two whole notes, and is prolonged through bar 16. The transformation of phrases two and three into six-bar units results in a succession of five phrases rather than six. This alteration disrupts the symmetrical pattern of whole notes that operates in the chorus of *That Mysterious Rag*.

Despite the asymmetrical phrasing of Satie's chorus, the choruses of the two rags sound much more similar than the verses, because Satie's chorus follows the prevailing diatonicism and the C major tonality of its model. His verse, on the other hand, contains a chromatic melody that bears little relation to the primarily diatonic writing of Berlin's verse. The chromaticism also disguises an important paraphrase in the verse—the duplication of Berlin's rhythms. Other marks of Satie's personal voice occur in his occasional use of melodic contours wholly unlike the contours of the Tin Pan Alley tune. In these instances, Satie tends to replace stepwise lines and repeated notes with triadic movement. The fifth

Ex. 5.12. Sketch of phrasing in the choruses of "That Mysterious Rag" and "Steamship Ragtime"

(a) "That Mysterious Rag"

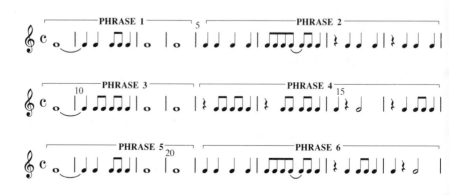

(b) "Steamship Ragtime"

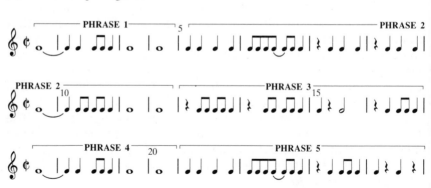

bar of the chorus of 'Steamship Ragtime', for example, substitutes a descending arpeggio for the repeated Cs of the Tin Pan Alley tune. Similarly, the penultimate bar of the chorus uses minor third movement from pitch A to pitch C in place of the stepwise turning figure of the Berlin model. With this greater frequency of triadic movement, Satie's 'Steamship Ragtime' acquires a raggier sound.

The paraphrase of *That Mysterious Rag* in *Parade* was radical in its use of a complete American popular tune in an art composition. It marks the only appearance in the ballet of a pre-existent melody; yet the tune sounds integrated rather than pasted on because Satie's paraphrase absorbs many of the same popular musical features—scalar and triadic writing, periodic phrasing, note-chord accompaniment—that characterize the other melodic types in *Parade*. In fact the melody in 'Steamship Ragtime' joins the ostinati, the sequential figures, and the melodic fragments as a fourth melodic type. It stands at the opposite pole from the ostinati, not because it introduces a new sound world, but because it occupies forty-eight bars and contains a wide range of pitches rather than three or four. Naturally Satie's incorporation of a Tin Pan Alley tune encouraged him to steer away from repeated accompaniment patterns and ostinati in favour of harmonic change, and to employ a primarily diatonic language. Yet the chromatic lines in the verse, and the pentatonic, whole-tone, and modal writing that frames the 'Steamship Ragtime' also make it one more voice among the many that comprise *Parade*.

MUSICAL DIVERSITY

Satie's choice of countless different melodic types in *Parade* and his continual introduction of new ostinati, new pendulum ideas, and new sequential figures leaves the listener encountering *Parade* for the first time with an impression of astonishing diversity. Short units follow one another in quick succession, the units tend to be only four bars long, and they often end abruptly without a cadence. If the ideas share a popular tunefulness, they contribute none the less to a highly diverse musical fabric because of their brevity and their abundance. The resultant loose structure is exactly what Satie intended. He wished to give his material a random quality and a sense of surprise and the unexpected in order to simulate the diversity and the juxtaposition typical of contemporary popular entertainment. It was precisely this kind of surprise that Apollinaire responded to in his programme note.

The music that opens the Little American Girl's section is composed in a whirlwind style that permeates much of *Parade* (see Ex. 5.6). An ostinato based on the simple pendulum principle cascades down at lower octaves for eight bars, whereupon it introduces the first ragtime theme of the Little American Girl scored for clarinet and tuba in B minor (bars 9–12). The theme lasts only four bars. It is followed by four bars of banal filler music—repeated two-note alternations in syncopated rhythms—before we move abruptly to whole-tone writing, scored idiomatically for strings and woodwinds (bars 17–24). Satie intentionally sharpens the contrast between the Impressionistic whole tone figures and their diatonic surroundings by preceding them with the brassy ragtime melody and following them with brassy B*b* major chords that convey a popular, syncopated style (bars 25–8). The result is a humorous spoof on Debussy's language. The whirlwind continues with a descending scalar idea and a bitonal clash between E*b* and A major scales (bars 29–34), before arriving on a long static ostinato.

Sections of *Parade* which move especially rapidly and suddenly from one idea to the next tend to contain shifts in metre, texture, timbre, and mode which reinforce the change in musical material. In bar 4 of Ex. 5.13, Satie breaks off the fugal episode of the previous three bars and inserts a descending scalar figure in harps and strings. The introduction of this figure is accompanied by a shift from counterpoint to monophony, and from C major to a Mixolydian mode on E. Satie heightens the diversity of sound, moreover, by

Ex. 5.13. Musical diversity in *Parade*

interrupting the descending scalar line after four bars and presenting first a two-bar (bars 8–9) and then a four-bar unit (bars 10–13), each distinguished by entirely different orchestrations. The first features lively brass and winds, while the second emphasizes a lighter piccolo, harp, and string texture. Satie reinforces the contrast by shifting the metre from duple to triple.

The element of surprise championed by Apollinaire is sometimes sparked by Satie's insertion of units that are less than four bars long. Satie assigns each of these units a different musical figure and a different number of bars. With two- and three-bar units whizzing by in rapid succession, and each unit differing in length from the previous one, the listener is unable to predict what change will come next and when it will occur.[18] Such passages intentionally avoid the simple tunefulness of the four-bar ostinati groupings that characterize the Chinese Magician's section. They highlight juxtaposition instead.

The many instances of musical diversity in *Parade* are enhanced by Satie's inclusion of a wide range of musical styles and languages within his twenty-minute ballet. Chorale, fugue, and Tin Pan Alley tune collide with modal, pentatonic, and whole-tone writing and with Satie's principal language of diatonicism peppered with occasional chromatic decoration. Parisians encountered a similarly eclectic sound world during their visits to circus, music-hall, and fairground in the early twentieth century.

MUSICAL NOSTALGIA

Satie's illustration of Cocteau's scenario extends to a remarkable evocation of the mood of nostalgia. The fairground setting and the three music-hall numbers in *Parade* were intended, in Cocteau's words, to 'represent varieties of nostalgia hitherto unknown'.[19] In the same vein, Cocteau associated his characters with an 'unromantic melancholy', a 'touching simplicity'[20] resulting from the contrast between the surface buoyancy of the music-hall acts and the wistfulness beneath. Auric's preface to the four-hand piano version of *Parade* similarly described Satie's evocation of 'the sorrow of the trestles,—the nostalgia of the primitive organ which will never play Bach fugues'. Such nostalgia had no precedent in Satie's earlier compositions. In this respect, as in the paraphrase of an entire American popular tune, *Parade* was unique for its time. It laid the foundation, moreover, for similar nostalgic utterances in two works by Francis Poulenc: the song cycle *Cocardes* (1919) and the ballet *Les Biches* (1924).

Satie's technique of creating a nostalgic, bittersweet tone consists principally of contrasting several bars of loud, resonant, richly

[18] For examples of a series of two- and three-bar units that create diversity of sound, see bars 414–29 and bars 494–510 of the Acrobats' section.
[19] See Jean Cocteau, *The Cock and the Harlequin* in *A Call to Order*, trans. Rollo Myers (New York: Haskell House, 1974), p. 25.
[20] Cocteau, *The Cock and the Harlequin*, p. 26.

textured music with a phrase or two of solo winds and strings playing
in a high register. The confident, brassy chorale of the opening, for
instance (see Ex. 5.14, bars 1–8), which simulates the lively
ballyhoos of the Managers, is followed by repeated plaintive
statements in solo flute and first violins marked 'very expressive'
(bars 9–11). The effect of these tenuous statements is to undermine
the chorale's bravado and to introduce a bittersweet hue. Towards
the end of the Chinese Magician's section, shown in Ex. 5.15,
descending stepwise figures marked 'plaintive' are repeated like
sighs in the solo oboe and violins, while the solo clarinet introduces
a listless sequential phrase. The nostalgia of this passage strikes the
listener particularly in retrospect, when Satie follows it with four
bars of lively, driving, syncopation (bars 9–11).

In the last bars of the ballet, Satie conveys the poignant failure of
the Managers and the three music-hall numbers to entice the crowd
inside (see Ex. 5.16). The final attempts of the Chinese Magician,
the Little American Girl, and the Acrobats are represented by a
repeated, emphatic pattern (see bars 1–2, B♭ clarinet, cornet)

Ex. 5.14. Nostalgia in *Parade*, Choral

Ex. 5.15. Nostalgia in *Parade*

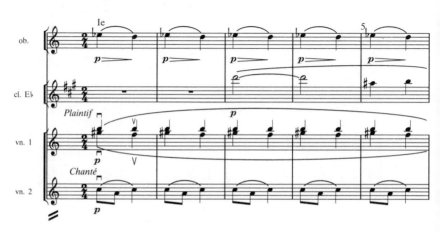

that grows increasingly aggressive as its tempo mounts. After eight bars the pattern breaks off suddenly, marking the collapse of the sideshow, and Satie inserts a few bars of the fugue from the opening of the ballet (bars 9–14). The fugue's hushed dynamics and high register give rise to a pathos that is especially stirring because it contrasts with the preceding frenzy and recalls the past and the high hopes that the Managers first entertained. With the return of the opening music, they are left facing the same predicament. Their hopes for success, nourished at the beginning of the ballet, have now vanished.

The 20 December 1920 issue of *Comoedia* contained a statement by Cocteau which used his earlier attacks on schools of musical composition as a springboard to separate *Parade* from labels of 'Dadaism':[21]

[21] Michel Sanouillet, *Francis Picabia et '391'* (Paris: Eric Losfeld, 1966), ii. 60.

Ex. 5.16. Nostalgia in *Parade*, Final

When we presented *Parade*, dadaism was unknown. We had never heard it spoken of. Now, there's no doubt that the public recognizes DADA in our well-meaning horse [the reference is to the third Manager—the two-man horse].

Well I like my friends Picabia and Tzara. If they were in need, I would lend assistance. BUT I AM NOT A DADAIST. Without doubt that is still the best way to be.

No, *Parade* is neither dadaist, nor cubist, nor futurist, nor of any school. *PARADE* IS *PARADE*, that is to say a big play-thing.

The description of Satie's ballet as a big toy flies somewhat in the face of Cocteau's insistence that *Parade* had no ties with the anti-academic, life-celebrating strains of Dada. More significant is his possibly unwitting prediction of what came to pass seven years after the première of *Parade*, when Satie's association with the Dada group triggered the rupture with his young disciples. Until then, the ballet's new popular sound world of simplicity, diversity, and nostalgia proved fertile ground for the music-making of Milhaud, Poulenc, and Auric.

6

Embracing a Popular Language

BUOYED by the success of *Parade*, Cocteau invented new ballets and musical events which fused 'art' and modern life. Three collaborative occasions, organized between 1918 and 1921, epitomize the successful merging of popular and concert-music worlds. The first occurred late in 1918, when Cocteau conceived of a 'Séance Music-Hall' and selected Francis Poulenc to collaborate with him on a circus piece, a satiric café-concert song, and a Prelude for percussion instruments. In February 1920 Cocteau devised his second event. This novel 'Spectacle-Concert' drew its title and programme from a typical evening at the Parisian music-hall and presented a new café-concert song by Poulenc, an instrumental parody by Satie, as well as dance music by Auric and Milhaud. The latter featured several famous clowns from the Nouveau Cirque and the Cirque Médrano. Finally, in 1921 the Swedish ballet director Rolf de Maré commissioned Cocteau and members of 'Les Six' to produce a 'pièce-ballet'. Cocteau contributed the scenario and choreography and five of 'Les Six' wrote musical numbers for *Les Mariés de la tour Eiffel* (The Eiffel Tower Wedding Party), a ballet with absurd dialogue and a strong reliance on music-hall revue, fairground games, and popular dance. In these songs, instrumental pieces, and ballets, and in their other collective and independent enterprises during the post-war years, Satie, Milhaud, Poulenc, and Auric made ample use of simultaneity, parody, simplicity, and repetition to challenge the loftiness of Romanticism.

MUSIC-HALL 'SPOOFS', CHILDREN'S TUNES AND FURNITURE MUSIC, 1918–1919

Cocteau sent Poulenc his proposal for the 'Séance Music-Hall' on 13 September 1918 and with his letter enclosed the words for a cafe-concert song, 'Toréador'.[1] He also requested a circus piece and a Prelude to introduce it. Poulenc complied and wrote all three pieces, but the Séance never occurred. The circus piece, which he

[1] See Francis Poulenc, *Correspondance 1915–1963*, ed. Hélène de Wendel (Paris: Éditions du Seuil, 1967), p. 21.

entitled *Jongleurs* or *Jugglers*, and the Prelude, scored for percussion, were later destroyed.[2] 'Toréador' remains, however, as evidence of Cocteau's unprecedented effort to organize a musical event around the theme of popular amusement. The commission of 'Toréador' was significant, moreover, because it prompted Poulenc's first experiments with an approach to songwriting that he employed throughout his career: the imitation of written features of French music-hall songs (syllabic setting, alteration of stress patterns) and aspects of the performance style of Maurice Chevalier (free treatment of the text).

Poulenc called 'Toréador' a ' "spoof" on café-concert song', and his implication—that the song both imitates and diverges from café-concert procedures—is apt.[3] The eight-bar introduction, the alternation between verse and refrain, the waltz tempo, and the use of an accompaniment in which low notes fall on strong beats and higher-register chords on weak beats place 'Toréador' squarely within the tradition of popular song. Yet Cocteau's text, and in more subtle ways Poulenc's music, also parody certain features of the French café-concert and music-hall repertoire. The use of parody reflects a taste for impiety and irreverence that characterized Parisian popular entertainment and comprised part of the appeal of popular spectacles for Poulenc and Cocteau. By selecting café-concert and music-hall as targets of satire, Poulenc and Cocteau create a situation in which popular amusement mocks itself.

Cocteau's words for 'Toréador' satirize the geographical incongruities of French music-hall songs. Just as the texts of early twentieth-century French songs often juxtapose different countries, describing, for example, a Frenchman in North Vietnam who studies the geography of China and Manchuria ('Petite Tonquinoise'), so Cocteau invents a story about a poor Spanish Toreador who is killed during a bullfight in the unlikely location of the square of St. Mark's in Venice.[4] Cocteau calls attention to this discordant mixture of Spanish and Italian cultures by referring successively to Spanish women, Venetian gondoliers, the doge of Venice, and the Toreador. He subtitles his song 'chanson hispano-italienne'.

---

[2] For this information on Poulenc's *Jongleurs* and Prélude, see Keith W. Daniel, *Francis Poulenc: His Artistic Development and Musical Style* (Ann Arbor: UMI Research Press, 1982), p. 17.

[3] Poulenc described 'Toréador' as 'une fausse chanson de caf'conç' in conversations with Claude Rostand. See Francis Poulenc, *Entretiens avec Claude Rostand* (Paris: René Julliard, 1954), p. 135.

[4] In his interviews with Rostand, Poulenc described the way in which 'Toréador' satirizes the geography of café-concert songs and cited an example from the French popular repertoire in which a Japanese takes a stroll in Peking. See Poulenc, *Entretiens*, p. 135.

I

Pépita reine de Venise
Quand tu vas sous ton mirador
Tous les gondoliers se disent:
Prends garde Toréador!
Sur ton cœur personne ne règne
Dans le grand palais où tu dors
Et près de toi la vieille duègne
guette le Toréador.
Toréador brave des braves
Lorsque sur la place Saint Marc
Le taureau en fureur qui bave
Tombe tué par ton poignard
Ce n'est pas l'orgueil qui caresse
Ton cœurs sous la baouta d'or
Car pour une jeune déesse
Tu brûles toréador.

    Belle Espagnole
    Dans ta gondole
    Tu caracoles
    Carmencita
    Sous ta mantille
    Œil qui pétille
    Bouche qui brille
    C'est Pépitaaa.

II

C'est demain jour de Saint Escure
Quand tu vas sous ton mirador
Le canal est plein de voitures
Fêtant le Toréador!
De Venise plus d'une belle
Palpite pour savoir ton sort
Mais tu méprises leurs dentelles
tu souffres Toréador.
Car ne voyant pas apparaître,
Caché derrière un oranger,
Pépita seule à sa fenêtre
Tu médites de te venger.
Sous ton caftan passe ta dague
La jalousie au cœur te mord
Et seul avec le bruit des vagues
Tu pleures toréador.

    Belle Espagnole . . .

III

Que de cavaliers! que de monde!
Remplit l'arène jusqu'au bord
On vient de cent lieues à la ronde
T'acclamer Toréador!
C'est fait il entre dans l'arène
Avec plus de flegme qu'un lord
Mais il peut avancer à peine
le pauvre Toréador.
Il ne reste à son rêve morne
Que de mourir sous tous les yeux
En sentant pénétrer des cornes
Dans son triste front soucieux
Car Pépita se montre assise
Offrant son regard et son corps
Au plus vieux doge de Venise
Et rit du toréador.
          Belle Espagnole . . .

(I) Pépita queen of Venice | when you go under your observation post | all the gondoliers say to each other: | Be careful, Toréador! | No one rules over your heart | in the big palace where you sleep | and next to you the old duenna | watches the Toréador. | Toréador, bravest of brave men | when on Saint Mark's Square | the bull in a fury which casts aspersion | falls, killed by your dagger | it is not pride which caresses | your heart under the baouta of gold | because for a young goddess | you burn, toréador. | Beautiful Spanish woman | in your gondola | you gambol | Carmencita under your mantilla | eye that sparkles | mouth that glitters | it's Pépitaaa. (II) It's tomorrow, Saint Escure's day | when you go under your observation post | the canal is full of cars | celebrating the Toréador! | From Venice more than one beautiful woman | quivers to know your fate | but you despise their lace | you suffer, Toréador. | For not seeing appear, | hidden behind an orange tree, | Pepita alone at her window | you meditate on avenging yourself. | Under your caftan slips your dagger | the jealousy of the heart gnaws at you | and alone with the noise of the waves | you cry toréador. | Beautiful Spanish woman . . . (III) So many horsemen! So many people! | fill the arena to its limits | people come within a radius of 100 miles | to acclaim you, Toréador! | It's done, he enters the arena | with more coolness than a lord | but he can scarcely advance | the poor Toréador. | There remains of his mournful dream | only to die beneath all the eyes | while feeling the horns penetrate | into his sad worried forehead | for Pepita appears seated | giving her glance and her body | to the oldest doge of Venice | and laughs at the toréador. | Beautiful Spanish woman . . . [Translated by Nancy Perloff]

Poulenc, for his part, creates satire through his treatment of the text. He employs a syllabic setting in the verse and in much of the refrain of 'Toréador', and in the final nineteen bars of the verse he highlights this syllabic writing by shortening note values, as shown in Ex. 6.1*a*. In this way, lines like 'Ce n'est pas l'orgeuil qui . . .' (bars 11–13) and 'Car pour une jeune . . .' (bars 19–21) rush by at

Ex. 6.1. Syllabic setting in "Toréador" and in a French music-hall song

(a) "Toréador"

(b) French music-hall song

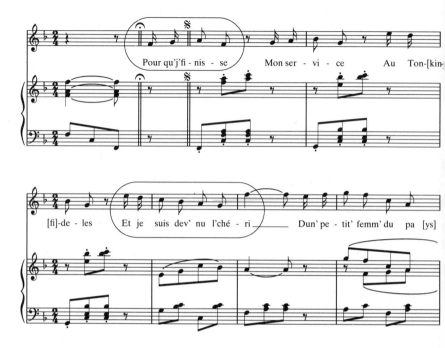

breakneck speed. Poulenc's technique of assigning a different syllable to each sixteenth note resembles the practice in French music-hall songs of piling several syllables on one pitch and creating a comical distortion of the text (see Ex. 6.1*b*). The distortion is exaggerated and parodied in 'Toréador', because the quick syllabic lines appear after a section that moves sedately in eighth and quarter notes (see Ex. 6.1*a*, bars 1–10).

Poulenc also parodies a technique found both in French music-hall songs (see Ex. 6.2*a*) and in art music of the period, of stressing words and syllables not normally accented in spoken French. Igor Stravinsky applied this technique in the first of his *Three Japanese Lyrics* (1913) which follows the model of the Russian *pribaoutki* (a form of popular Russian verse) by ignoring the accents of the spoken verse when the text is sung. In the refrain of 'Toréador', Poulenc's emphasis on weak beats consists of creating sudden stops on the last 'e' of each line of text, as shown in Ex. 6.2*b*, and calling attention to the 'e' with a grace note. This alteration shifts the stress patterns of spoken French, while at the same time creating an accurate if exaggerated portrayal of the Toreador's Spanish language

Ex. 6.2. Alteration of stress patterns in a French music-hall song and in "Toréador"

(a) French music-hall song

(b) "Toréador"

which stresses the final 'e' and gives it a full vowel sound.[5] The use of the grace note can be traced directly to recent fashions in art music: grace notes in the chamber accompaniment to the first of Stravinsky's Japanese songs highlight the accentuation of weak syllables in the Russian text.[6]

Poulenc's approach to prosody in 'Toréador' is an intriguing blend of music-hall techniques, art music tastes, and the performance style of his idol, Maurice Chevalier. In a statement that appeared in *Conferencia-Les Annales* in December 1947, Poulenc acknowledged his debt to Chevalier:[7]

You will sense immediately the difference between my songs and my *mélodies*. Indeed the word song implies, in my eyes, a style that without belonging to any folklore, does not mean to any lesser extent a total liberty with regard to the text. I begin words again, I cut them up, I even create ellipsis . . . The *tour de chant* of Maurice Chevalier has taught me much in this respect.

A Pathé recording from the 1910s featuring Chevalier's performance of the song 'Si fatigué' illustrates the free treatment of the text described by Poulenc. Claude Rostand notes that Poulenc owned the recording and played it to demonstrate the impact of Chevalier's prosody on his own approach to text setting.[8] On this recording and on others of the period, Chevalier creates rubato by elongating syllables or delaying his attack so that the melody lags behind the accompaniment.[9] Occasionally he speaks rather than sings his lines, and at these junctures he departs from the steady pulse, recovering the original pace after a few phrases. In the refrain he splits up words, as in the line 'J'étais si fa, j'étais si fa, j'étais si fatigué'. Taking liberty with the text produces a striking freedom from adherence to a regular metre and tempo.

[5] Stereotyped sounds of Spain are also evoked and exaggerated at the end of the refrain where melodic phrases in Phrygian mode, the mode of Spanish gypsy music, recur identically for several bars. For a recording which illustrates the sudden stops on the final 'e' of each bar in the refrain, see *Francis Poulenc Mélodies* (Elly Ameling, Nicolai Gedda, William Parker, Michel Sénéchal, Gérard Souzay; Dalton Baldwin, piano), EMI 2C 165-16231/5.

[6] Stravinsky's composition of the *Japanese Lyrics* was inspired by his discovery of an anthology of Japanese verse and his selection of three lyrics for translation into Russian by A. Brandta. English, German, and French translations were also made, but Stravinsky insisted that he would rather not hear his Russian songs at all than hear them sung in translation. He wished the mis-accentuation to be heard.

[7] This statement is quoted by Keith Daniel in *Francis Poulenc*, p. 244. Translations here and throughout the chapter are by the present author unless otherwise indicated.

[8] See Francis Poulenc, *Entretiens avec Claude Rostand*, p. 136.

[9] The description of Chevalier's performance style is based on the present writer's observations after hearing the recording of 'Si fatigué' (Pathé Saphir 4181) at the Phonothèque Nationale in Paris. Additional song performances by Chevalier are available on *La Fete du Caf'Conc' 1900–1925*, Pathé Marconi 2 C 150 15370/71M.

In 'Toréador' we find an exaggerated treatment of the techniques of Chevalier. Whereas Chevalier manipulates text and tempo primarily in the refrain, Poulenc uses both verse and refrain to explore declamatory setting and elastic tempo. Ex. 6.3, taken from the verse, illustrates how Poulenc shifts the metre from triple to duple and introduces short note values that are sung rubato (bars 4–5) in order to interject Chevalier's fluid singing style. Towards the end of the verse Poulenc does not alter the metre, but his sudden use of a melodic line of sixteenth notes set syllabically and followed by bars of rest, and his substitution of an asymmetrical chordal pattern for the regular, lilting accompaniment of the previous bars exaggerates and satirizes Chevalier's declamatory style (see Ex. 6.1*a*, bars 11–21). Throughout 'Toréador', Poulenc pays special homage to Chevalier's penchant for breaking into speech in the midst of song, by placing a new word or syllable on each successive pitch, no matter how short the note values.

Ex. 6.3. Elastic tempo in "Toréador"

Poulenc's colleague Georges Auric, who set verses of Cocteau to music in the summer of 1918, just months before 'Toréador', also parodied the performance style of French music-hall singers. Auric's *Huit Poèmes de Cocteau* (voice and piano) were not commissioned for a music-hall event. Yet the songs contain striking fluctuations in tempo and metre which suggest the influence both of Chevalier and of two other French music-hall singers, Mayol and Mistinguett, who made liberal use of rubato in their performances. Auric establishes the context of popular song in *Huit Poèmes* by selecting a steady tempo and a clear pulse for the opening bars of each piece. After one or more phrases, he begins shifting the tempo and introducing a speech-like fluidity. In 'Place des Invalides', for example, shifts in tempo and in metre occur so frequently that Auric's imitation of music-hall performance practice takes the form of caricature (see Ex. 6.4).

Varying tempos and metres in the *Huit Poèmes de Cocteau* occur in the context of startling juxtapositions of musical material. Auric achieves such oppositions by presenting a series of short tuneful

Ex. 6.4. Shifts in tempo and metre in "Place des Invalides"

ideas, separated from one another by rests or piano interludes and distinguished by tonality, rhythm, and tempo. In the middle of the first song, 'Hommage à Erik Satie' (see Ex. 6.5, bar 1), we hear a tune with leaps of thirds, fourths, and fifths suggestive of a fanfare and an accompaniment that follows the rhythms of a 'marche militaire'. The theme stands apart from its surrounding material, not only because of its martial quality, but because it opens and closes in a new key, C major, and returns to the allegretto tempo that began the song. Auric heightens the tune's independence by juxtaposing it against a piano interlude and then a contrasting idea in E major in the voice (bar 18).

Both the fanfare and the E major fragment are typical of the separate tunes that comprise the fabric of 'Hommage à Erik Satie' and all the songs in *Huit Poèmes*. Other independent tunes in the set imitate the simple, repetitive style of a French children's ditty.[10] The abundance of these melodies combined with the frequent and abrupt changes in accompaniment figures confronts the listener with a startling simultaneity. Indeed, listening to *Huit Poèmes* resembles a walk through the Paris streets, where one or two phrases resound from the many different popular milieux: fairground with its mechanical organs, music-hall with its songs and revue tunes, circus with its brassy music accompanying the clowns and the acrobats.[11]

Musical parody was rarely far from Georges Auric's mind. In *Huit Poèmes*, he uses the device of repetition to poke sharply and playfully at the Romantic credo. From his mosaic of accompaniment patterns, he selects two- and three-note figures and repeats them several times, without applying a clear function such as transition, introduction, conclusion, or ostinato. The resultant banality and humour create a deliberate 'spoof' on the perpetual variation, the shifting harmonies, and the general complexity and seriousness of nineteenth-century music.[12]

In addition to their parody and their imaginative reflection of the Parisian urban environment, Auric's *Huit Poèmes* mark one of the first attempts by a European art composer to capture the sound of

[10] Auric's choice of a style that imitates children's tunes pays homage to Satie's set of children's pieces, the *Enfantines* (1913) which includes *Menus Propos enfantins*, *Enfantillages pittoresques,* and *Peccadilles importunes*. In bar 31 of 'Hommage à Erik Satie', moreover, Auric quotes the first four pitches of Satie's *Trois morceaux en forme de poire.*

[11] For additional comments on the juxtaposition of 'commonplace melodies' in Auric's *Huit Poèmes de Cocteau*, see Paul Collaer, 'I "Sei": studio dell'evoluzione della musica francese dal 1917 al 1924', *L'Approdo musicale* 19–20 (1965): 41.

[12] A later collection of songs by Auric, *Alphabet* (1920), combines an imitation of folk and children's tunes with a satire of such Impressionist clichés as Spanish rhythms, quartal harmonies, and parallel movement.

Ex. 6.5. Juxtapositions of musical material in "Hommage à Satie"

lowered third and seventh scalar degrees, known as 'blue notes'.[13]
Auric's knowledge of 'blue notes' at such an early date must have
arisen from two circumstances: his presence at performances by
Louis Mitchell's Jazz Kings beginning in December 1917 at the
Casino de Paris; his likely attendance of concerts given by Jim
Europe's Hellfighters during their tour of France between 12
February and 29 March 1918. The repertoire of Mitchell's Jazz
Kings included the tune *Everybody Step* which makes occasional use
of lowered thirds. The Hellfighters performed and later recorded
W. C. Handy's *Memphis Blues* which contains instances of 'blue
notes' in the first of its three strains.[14] Although written 'blue notes'
do not abound in this repertoire, they may well have been
interpolated into the performances.

Armed with a listener's knowledge of lowered thirds, Georges
Auric employed them discreetly in the piano accompaniment of
Huit Poèmes. In Ex. 6.6, the attempt to evoke a jazz sound is

Ex. 6.6. Suggestion of "blue notes" in *Huit Poèmes de Cocteau*

apparent from Auric's use of a sliding movement between the raised
second and scale degree three. One must stop short of identifying
the raised seconds as 'blue notes', however, because although
enharmonically equivalent to the flattened 'blue notes', the seconds
in *Huit Poèmes* do not follow the characteristic blues resolution

[13] The only European concert work that introduced suggestions of 'blue notes' contemporaneously with Auric's *Huit Poèmes* is Stravinsky's *Ragtime* for eleven instruments. Barbara Heyman explains that Stravinsky completed the first sketches for *Ragtime* in November 1917. See Barbara Heyman, 'Stravinsky and Ragtime', *The Musical Quarterly* 68, No. 4 (Oct. 1982): 547. According to Eric Walter White, the instrumental score of *Ragtime* was finished on 11 November 1918. (Auric completed *Huit Poèmes* in summer 1918.) See Eric Walter White, *Stravinsky: The Composer and His Works* (Berkeley: University of California Press, 1979), p. 275.

[14] The repertoire of Mitchell's Jazz Kings can be gleaned from recordings made by Mitchell's group between 1921 and 1923 on Pathé 6544, 6550, 6574, 6599, 6607. These recordings are included on *Paris, 1919–1923, premiers jazz-bands*, Pathé Marconi 172 7251. The Hellfighters' recordings of *Memphis Blues* (Pathé 22085-mx No. 67486) and *Clarinet Marmelade* (Pathé 22167-mx No. 67668) appear on *Steppin' on the Gas: Rags to Jazz 1913–1927*, New World Records 269. Sheet music of this repertoire is available in Special Collections, New York Public Library, and at the Library of Congress.

down to scale degree one. Auric's spelling of these pitches as raised rather than lowered degrees, moreover, identifies them as alterations of scale degree two, hence strengthening the link with concert music. Yet Auric believed, perhaps, that he was imitating a jazz-blues usage.

More than a year after Auric's composition of *Huit Poèmes*, Darius Milhaud wrote music to accompany poetry of Cocteau. Milhaud's *Trois Poèmes de Jean Cocteau*, for voice and piano, were composed between October and December 1919 following his return from Brazil, and may have been inspired by the settings of Poulenc and Auric. A salute to the new French school is apparent in Milhaud's choice of verses by Cocteau and his dedication to Satie. The songs acknowledge the popular basis of the new sensibility by referring to various aspects of Parisian popular entertainment. 'Fumée' ('Smoke') describes the horseback rider jumping through hoops at the Cirque Médrano. 'Fête de Bordeaux' (Bordeaux Festival) speaks of the merry-go-round at the fair, and 'Fête de Montmartre' evokes night-time boats and sailors.

For his setting, Milhaud pays homage to Satie's children's pieces of 1913, the *Enfantines*, and to his piano pieces with folk song quotations.[15] The simplicity of folk and children's tunes offered Milhaud and his associates a useful alternative to the complex melodic and harmonic writing of Romanticism. In 'Fumée' the piano accompaniment, shown in Ex. 6.7, opens and closes with a perfect fifth (B–F♯) sounding in the bass, in imitation of the drones

Ex. 6.7. Folk elements in "Fumée"

[15] Examples of Satie's piano pieces with folk song quotations include *Vieux Sequins et vieilles cuirasses* (1914) which paraphrases 'Malbrouck s'en va t'en guerre' and 'Le Roi Dagobert', and *Sur un vaisseau* from *Descriptions automatiques* (1913) which incorporates 'Maman les petits bateaux'. Satie quotes 'Au claire de la lune, mon ami Pierrot' (which was also used by Debussy in his early song 'Pierrot' (1881)) in *The Flirt* from *Sports et divertissements*.

that accompany French folk tunes, and a simple pentatonic melody
heard above. 'Fête de Bordeaux' includes a four-bar tune in the
vocal line that bears a striking resemblance to French children's
songs, particularly to the well-known 'Il était une bergère' ('There
was a shepherdess'), in its opening upbeat from G–C (5–1) followed
by ascending and descending stepwise movement (see Ex. 6.8a–b).
In 'Fête de Montmartre', the final bars introduce four-bar phrasing,
melodic and rhythmic repetition within each phrase, and scalar and
triadic movement based exclusively on the pitches of the principal
tonality, Ionian on A. Such devices typify French children's music,
but are not unique to it. They also characterize the French
vernacular repertoire of songs, marches, and dances.

Ex. 6.8. "Fête de Bordeaux" and folk model

(a) "Fête de Bordeaux"

(b) "Il était une bergère"

The practice, seen in Milhaud's 'Fête de Montmartre', of creating a simple four- or eight-bar tune by compiling stylistic traits of folk and popular genres also characterizes Poulenc's 'Le Dauphin' ('The Dolphin'), from the collection *Le Bestiaire* (1918–19). The song, shown in Ex. 6.9, features a simple periodic theme in A major (Ionian). Alongside the tonal centring and the periodicity which

Ex. 6.9. Folk elements in "Le Dauphin"

typify both children's songs and popular tunes, Poulenc incorporates certain traits identified specifically with marches and others associated with children's tunes. The duple metre and the reiteration of perfect fourths (bars 2–3, 5–7) come from marches. The local repetition of pitch patterns (bars 2–3, 6–7) and the exact recall in phrase two of the rhythms of phrase one, convey the simplicity of children's music.

In a dramatic example of the interest art composers in pre- and post-war France shared in writing children's pieces, Poulenc modelled 'Le Dauphin' on two children's sets from 1913: Satie's *Enfantillages pittoresques* [Picturesque Child's-Play], from the *Enfantines*, and Stravinsky's *Three Little Songs*. The melody in the introduction to 'Le Dauphin' paraphrases both the principal tune of Satie's 'Marche du grand escalier' from *Enfantillages pittoresques* and the opening line of Stravinsky's 'The Jackdaw' from *Three Little Songs* (see Ex. 6.10*a–c*). Poulenc's melody borrows the rhythm of

Ex. 6.10. Melodic models for the Introduction to "Le Dauphin"

(a) "Le Dauphin"

(b) "Marche du Grand Escalier."

(c) "The Jackdaw"

four quarter notes followed by eighths and the contour of an opening stepwise descent followed by a turning figure. In addition, Poulenc's linear writing and sparse texture in 'Le Dauphin' can be traced to the pieces of Satie.

Another device clearly associated by Satie and Stravinsky with the evocation of children's music was the repetition of short units. Repeated two- and four-bar units in Satie's *Menus Propos enfantins* [Childish Small Talk] (1913) and Stravinsky's *Three Little Songs* inspired a similar kind of repetition or varied restatement in Poulenc's important set of piano pieces from 1918, *Mouvements perpétuels*. The continuous repetition, the prominent melodic line, the thin texture, and the frequent ostinato accompaniments indicate that Poulenc was intrigued both by childlike re-creations and by the sounds of banality and repetition. His published score of *Mouvements perpétuels* (1919) contains such indications as 'colourless', 'without nuances', and 'indifferent'. In the original manuscript (1918), moreover, he emphasizes the conception of banal, monotonous music by placing 'reprise ad libitum' at the end of each piece and preceding the set with the following notice:[16] 'These three pieces ought to be performed only so as to be linked one with the other and closely linked. The whole should unroll uniformly and in a colourless manner. The pianist ought to forget that he is a virtuoso.'

Poulenc omitted this notice in the published version, along with the performance directions calling for a 'reprise ad libitum'. Yet the manuscript of *Mouvements perpétuels* makes it clear that Poulenc intended these pieces to convey the monotony of background music. His conception was inspired by Erik Satie: as early as 1914, Satie included in his *Sports et divertissements* a 'Perpetuel Tango' which instructed the performer to go back to the beginning and repeat the dance over and over again. Satie's symphonic drama, *Socrate*, completed in the same year as Poulenc's *Mouvements* (1918), called for 'musique d'ameublement' to accompany the banquet scene.[17] Two years later, on 8 March 1920, Satie and Milhaud introduced the idea of background music to the public when they organized a concert at the Galerie Barbazanges in which 'furnishing music', made up of popular refrains, was played by a small band during the intermissions. Although Poulenc did not participate in this event,

---

[16] The 1918 holograph of Poulenc's *Mouvements perpétuels* that contains this notice and the indications for 'reprise ad libitum' is available in Special Collections at the New York Public Library.

[17] Satie first used the term when he was jotting down plans for *Socrate* in one of his notebooks. The jottings are included in Pierre-Daniel Templier, *Erik Satie* (Paris: Rieder, 1932), p. 46.

the manuscript of *Mouvements perpétuels* reveals that he joined Satie in experimenting with the concept of furniture music two years before the concert at the Galerie Barbazanges.

For Poulenc, Satie, and Milhaud, the creation of simple repetitive music that placed few technical demands on the performer was a means of defying the academic world of concert music. Repetition also implied circularity, a lack of progression or development, a playful stance, all of which flew in the face of the Romantic conception of art music.

THE 'SPECTACLE-CONCERT' OF 1920: A MUSIC-HALL RE-CREATION

Following his abortive attempt at a 'Séance Music-Hall', Cocteau successfully organized a new event in February 1920—a 'Spectacle-Concert'—similar in concept to the projected 'Séance'. Four compositions, one by Satie and one by each of the younger composers, received premières.[18] In his autobiography, Georges Auric reports that he and Satie wrote their pieces especially for the occasion. Satie, in particular, was so 'seduced by this next adventure' that he finished his work well in advance of the concert date.[19] According to Auric, the performance of the four pieces at an important Parisian theatre, the Théâtre de Comédie des Champs-Élysées, contributed significantly to public and critical recognition of the group of French composers. Before the 'Spectacle-Concert', programmes featuring music by Satie and his disciples also included works of minor French composers such as Roger-Ducasse and Roland-Manuel and usually took place at a small artist's studio, the Salle Huyghens. The 'Spectacle-Concert', however, offered for the first time exclusive devotion to music of Satie, Milhaud, Poulenc, and Auric at an established Parisian performance venue.

Cocteau's unique conception strengthened the significance and novelty of the event. Art compositions based on popular idioms were selected for performance at a re-created music-hall evening. The intention was to simulate the blend of the aural and the visual, of song, dance, and theatre, as well as the dazzling variety and simultaneity of music-hall entertainment. Cocteau's approach to the event reflects his close adherence to the popular model. He entitled his evening 'Spectacle-Concert', after a term commonly seen on posters of the period to advertise café-concert and music-hall shows (see Plate 9).

[18] A fifth work, an *Ouverture* by Poulenc, also received its première. The *Ouverture* offered only a brief introduction to the programme, however, and comprised an excerpt from a longer work, Poulenc's *Sonate* (1918) which featured a witty *mélange* of the styles of Stravinsky and Satie from the early 1910s.

[19] See Georges Auric, *Quand j'étais là* (Paris: Bernard Grasset, 1979), p. 158.

His programme mirrored the typical music-hall evening by including clown and acrobat performers and by dividing the evening into a series of parts ('parties'), one for dances, a second for songs and instrumental divertissement, and a third for a play or short revue.[20] Auric's foxtrot, *Adieu, New York!*, occupied the dance portion and featured the clowns Foottit and Jackly in an acrobats' dance organized by Cocteau. Poulenc's *tour de chants, Cocardes* (Cockades), and Satie's set of instrumental pieces, *Trois Petites Pièces montées*, corresponded to the *partie-concert* or musical section of the evening.[21] Finally, Milhaud's musical pantomime, *Le Bœuf sur le toit* (The Ox on the Roof), was a blend of play and revue. Elements of theatre appeared most prominently in the use of one plot, rather than a series of different skits, and in the choice of eight characters. The revue was a model in the selection of the Fratellinis and other celebrated clowns from the Cirque Médrano to perform the roles, and in the addition of Milhaud's orchestral music.

As a final element of authenticity, Cocteau set up an 'American bar' at intermission and hired musicians to perform.[22] The bar occupied one part in a complex array of circus performers, music-hall acts, and art compositions that went into the making of the 'Spectacle-Concert'. The result was a radical merging of popular and concert-music arenas.

Cocteau's event epitomized the 'esprit nouveau' or new spirit identified and publicized by Apollinaire in his note on *Parade* in May 1917 and in a longer lecture, 'L'Esprit nouveau et les poètes', delivered in November 1917.[23] In both statements, Apollinaire called for the renunciation of beauty, of 'formulas', of Wagner's

[20] For a description of the different 'parties' featured on a music-hall programme at the Scala, see ch. 2. Jacques Feschotte discusses the typical format of music-hall programmes in *Histoire du music-hall* (Paris: Presses universitaires de France, 1965), p. 27.

[21] Poulenc's *Cocardes* is described as a *tour de chant* on the original programme of the 'Spectacle-Concert', displayed in the exhibition 'Cocteau Generations: Spirit of the French Avant-Garde' at The Grey Art Gallery and Study Centre, New York University, 17 May–23 June 1984.

[22] A brief description of the mandolin players at the bar appears in Louis Laloy, 'À la Comédie des Champs-Elysées', *Comoedia* (23 Feb. 1920), p. 1, and in Francis Steegmuller's *Cocteau: A Biography* (Boston: Little, Brown, & Co., 1970), p. 243. The concert programme described the bar as an 'American bar'.

[23] Apollinaire's note on *Parade* first appeared in the newspaper *Excelsior* on 11 May 1917 and then a week later in the ballet programme. The statement, which contained the first use of the term 'sur-réalisme', is translated in Steegmuller, *Cocteau*, p. 512. Steegmuller explains that Count Étienne de Beaumont financed the 'Spectacle-Concert' in the hope of interesting the public in *l'esprit nouveau*. See Steegmuller, p. 242. Richard Axsom discusses the shared interests of Cocteau and Apollinaire in an *esprit nouveau* in *'Parade': Cubism as Theater* (New York: Garland Publishing Inc., 1979), pp. 53–4. Apollinaire delivered the essay as a lecture in late 1917 at the Théâtre du Vieux-Colombier; it was first published on 1 December 1918 in *Mercure de France*.

'heavy romanticism' in favour of truth, novelty, contradiction, and surprise. 'Poets are not only men of beauty,' Apollinaire wrote. 'They are still and above all men of truth, in so far as truth enables them to penetrate the unknown, as well as the surprise, the unexpected.'[24] Such concerns were central to the aesthetic of the French and Italian avant-garde. The expression of the fantasy and the 'magic of everyday life' which characterized *Parade* was recast in the 'Spectacle-Concert'.[25]

In a preface to this event, Cocteau voiced an attack on Romanticism similar to Apollinaire's and praised instead the 'musique à l'emporte-pièce' (trenchant music) which Satie and his colleagues were presenting on the new programme.[26] The slogan 'trenchant music' indicated the simple, direct quality of popular sources and stood at the opposite pole from 'musique à l'estompe', or the hazy language of Impressionism. Each of the four pieces on the programme explored different aspects of popular entertainment. *Adieu, New York!* made use of American syncopated dance styles; *Cocardes* imitated the performance practice of Poulenc's favourite music-hall singer, Maurice Chevalier; *Trois Petites Pièces montées* presented a parody of art music; *Le Bœuf sur le toit* re-created popular Brazilian sambas and maxixes and the simultaneity of fairground and music-hall.

Following an *Ouverture* which comprised the final movement of Francis Poulenc's *Sonate* for four-hand piano (1918), the programme opened with Georges Auric's foxtrot *Adieu, New York!* (1919). In its references to the everyday, the piece synthesizes the many different American popular styles that Auric encountered. From foxtrots such as those recorded in Paris by the black drummer Louis Mitchell and his Jazz Kings and from post-1911 rags available in Paris on Gramophone Records, he borrowed the formal device of successive sections or strains. He also employed the dotted rhythms (dotted eighth followed by sixteenth) that pervaded American foxtrots, ragtime tunes of the 1910s like E. Claypoole's *Ragging the Scale*, and Tin Pan Alley songs from *Alexander's Ragtime Band* to *The Darktown Strutters' Ball*.[27] A repeated two-bar figure that introduces the first strain (see Ex. 6.11a) imitates vamps in American

---

[24] See Apollinaire, 'L'Esprit nouveau et les poètes', p. 488.

[25] Apollinaire spoke of the 'magic of everyday life' conveyed in *Parade* in his note of 11 May 1917.

[26] Cocteau's preface to the 'Spectacle-Concert' is included in an article by Jean Marnold, 'Un Spectacle d'avant-garde: avant *Le Bœuf sur le toit* de M. Jean Cocteau', *Comoedia* (21 Feb. 1920), p. 1.

[27] For information on Gramophone recordings of American ragtime tunes, see Chapter 2. For Louis Mitchell's recording of the foxtrot *Everybody Step* in January 1922 (Pathé 6574),

Ex. 6.11. Milhaud's paraphrase of Auric's *Adieu, New-York!*

(a) Auric, *Adieu, New-York!* (fox-trot)

(b) Milhaud, *Caramel mou* (shimmy)

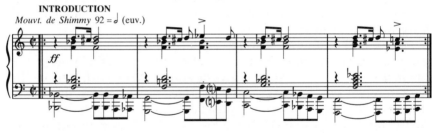

popular songs and dances. Auric's pooling of these formal, rhythmic, and motivic devices from several different American sources rather than exclusively from foxtrots creates a piece that Cocteau called a portrait of a foxtrot. Even within his foxtrot model,

see *Paris 1919–1923 premiers jazz bands*, Pathé Marconi 172 7251. *Alexander's Ragtime Band* and other Tin Pan Alley tunes were introduced to Parisians by the American dance team Vernon and Irene Castle. Edward Berlin explains the stylistic relationship between foxtrots, rags of the 1910s, and blues pieces in *Ragtime: A Musical and Cultural History* (Berkeley: University of California Press, 1980), pp. 147–52.

moreover, Auric transforms the typical practice of successive eight-
or sixteen-bar strains by extending or compressing the strain's
length and inserting the opening vamp into the middle of a strain.

The exclamation Auric selected as the title for his foxtrot was
completed by the words, 'Bonjour, Paris!', appearing above an
article written in May 1920 for *Le Coq*. In this article, Auric argued
that jazz had served its purpose. By offering a fresh model for
composers, it had 'dispersed the heavy seductions of Debussyism'
and revitalized French concert music. Now, Auric warned, it was
time to bid New York farewell and turn to French musical sources.[28]
The nationalistic message of Auric's article suggests that in 1919
when he composed *Adieu, New York!*, he intended the piece to be a
final tribute to jazz, a work that would close the period of American
influence. Auric set an example, moreover, by ceasing to incorporate
'blue notes', strains, and syncopation.

Auric's colleagues did not necessarily heed his call. Darius
Milhaud found the idea of a foxtrot so intriguing that he transcribed
Adieu, New York! for four-hand piano in December 1919, right
after Auric completed the piece. Then in April 1921 Milhaud wrote
his own version of an American dance, entitled *Caramel mou* (Soft
Caramel), a shimmy. Milhaud scored the work for jazz band and
dedicated it to Auric; the text for a vocal part was provided by
Cocteau. Just as *Adieu, New York!* had its première on the eclectic
programme of the Spectacle-Concert, *Caramel mou* first appeared
on a diverse programme of plays and incidental music. Just as
Adieu, New York! represented the dance portion, *Caramel mou* was
performed as musical accompaniment to a dance, this time featuring
the Black entertainer Gratton.[29] The choice of a Black dancer was
appropriate since the shimmy originated as a Black-American dance.
Popular in America during the 1910s and 1920s, the shimmy was a fast
version of the foxtrot. Milhaud's choice confirms that he wished his
piece to be a tribute to Auric's.

The two dances by Milhaud and Auric share several musical
traits. *Caramel mou* begins with a repeated four-bar introduction
similar in function to the two-bar vamp at the beginning of *Adieu,*
New York! The melody in Milhaud's opening strain paraphrases the

---

[28] The summary of Auric's argument here is culled both from the article 'Bonjour Paris!'
and from an article entitled 'Après la pluie le beau temps' which appeared in *Le Coq*, June
1920.

[29] Milhaud praised the diversity of the programme which included a play by Max Jacob, a
one-act play by Raymond Radiguet (*Le Pelican*) with music by Auric, a play by Cocteau and
Raymond Radiguet (*Le Gendarme incompris*) with music by Poulenc, and Satie's *Le Piège de*
Méduse. Pierre Bertin organized this 'avant-garde show' which took place at the Théâtre des
Mathurins on 23 May 1921. See Milhaud, *Notes without Music*, trans. Donald Evans, ed.
Rollo Myers (London: Dennis Dobson, 1952), pp. 103–4.

opening measures of the melody in the first strain of *Adieu, New York!*, as shown in Ex. 6.11*a–b*. In his treatment of rhythm, Milhaud follows Auric's example by selecting dotted figures. The use of dotted rhythms coupled with Milhaud's frequent application of blue notes reflect the presence of the same foxtrot, ragtime, and Tin Pan Alley sources that inspired Auric.[30]

The uniqueness of *Caramel mou* lies in its scoring for 'orchestre de jazz'. Milhaud selected an ensemble of clarinet, trumpet, trombone, saxophone, piano, percussion (drum, cymbals, bell, sticks), and voice ad libitum. This instrumentation closely resembles the seven-piece group that comprised Louis Mitchell's Jazz Kings. Milhaud's call for a singer and dancer suggests that he may also have been influenced by Mitchell's earlier band, the Seven Spades, which performed a three-week engagement in Paris sometime between 1915 and 1917 and featured a rag singer (Set Jones) and a dancer (F. Jones).[31] Another American group, the Scrap Iron Jazzerinos, made up of trumpet, trombone, saxophone, piano, banjo, whistle, and drums, may have provided an additional model for the instrumentation of *Caramel mou*. In February 1919, the Jazzerinos recorded a tune on Pathé called *Everybody Shimmies Now* which contains the dotted rhythms, the 'blue notes', and the loose sectional design adopted in *Caramel mou*. Hence the dance tunes of the Jazzerinos, as well as the group's instrumentation, offered a likely prototype.

In orchestrating *Caramel mou*, Milhaud followed the standard practice of the early jazz bands he had encountered. The trumpet takes the lead more or less consistently, while clarinet and saxophone shift between lead and countermelodies, and trombone plays both countermelodies and occasional bass line support.

Both *Adieu, New York!* and *Caramel mou* contain dissonance, chromaticism, passages of dense texture and, in Milhaud's piece, bitonality—all of which distinguish them from popular dances. Yet the sweeping use of American popular material in a concert piece was a gesture against romanticism. Furthermore, in *Caramel mou* Milhaud parodies serious music through the aimless repetition of banal musical ideas. In the melody of the opening strain, shown in

---

[30] Milhaud may also have been inspired by the dotted figures in Zez Confrey's famous novelty-piano piece *Kitten on the Keys*, which he mentions in *Études* (Paris: Éditions Claude Aveline, 1927), p. 55. *Kitten on the Keys* dates from 1921.

[31] Mitchell's Jazz Kings featured Mitchell on drums, Crickett Smith on trumpet, Joe Meyers on guitar, Dan Parish on piano, Walter Kildare on bass, Frank Withers on trombone, James Shaw on bass. For this information, as well as material on Mitchell's Seven Spades, see Robert Goffin, *Jazz: From the Congo to the Metropolitan*, trans. Walter Schaap and Leonard G. Feather (New York: Da Capo Press, 1975), pp. 58, 68.

Ex. 6.12, he repeats the pitches B*b*, C, D in ascending and descending order and applies subtle rhythmic variations. The repetition of the figure lasts for six bars and serves no other function than to arbitrarily extend the strain. The listener's unsuccessful attempts to predict when the three-note idea will cease produce moments of humour both then and later in the dance. Milhaud clearly enjoyed the contrast between his devices to thwart listener expectations (repetition, monotony), and the dramatic techniques of harmonic and formal disruption used by Beethoven to achieve the same ends of surprise.

Ex. 6.12. Banality in *Caramel mou*

Francis Poulenc did not share Milhaud's absorption with American syncopated dance music. Rather, his natural preferences lay with French sources and, as a result, he did not need Auric's admonitions to bid New York farewell. The song set *Cocardes* (June 1919), a group of three 'chansons populaires' which Poulenc dedicated to Auric, was the third composition on the programme of the Spectacle-Concert and the start of the evening's 'musical and song portion'. In a letter to Paul Collaer on 21 June 1920, Poulenc explained that his songs captured the essence of Paris and did so in a coarse, rough-hewn manner, 'without artifice'. 'They will clearly show you,' he wrote, 'that I am not an Impressionist.'[32] The texts of the songs, written by Cocteau, abound in references to circus, fair, cinema, and music-hall. Poulenc's music correspondingly employs three devices to portray the Paris streets as he knew them: first, a scoring for small band (violin, cornet, trombone, bass drum,

[32] The letter appears in Collaer, 'I "Sei"', *L'Approdo musicale* 19–20 (1965): 45, 47.

triangle, cymbal) in imitation of the Parisian bands performing at fairground sideshows and at dance-halls known as *bals-musettes*; second, an interpretation of the performance style of Maurice Chevalier; third, an expression of nostalgia which Poulenc tended to associate with the popular world.[33] The textual references to popular entertainment and the band instrumentation are obvious methods of depicting popular entertainment. The borrowing from Chevalier and the expression of nostalgia are more subtle.

Each of Cocteau's texts in *Cocardes* presents a kaleidoscopic succession of popular images: 'caramels mous', 'bonbons', 'trapèze', 'girafe', 'air de Mayol', 'cinéma'. The impression that words are selected randomly stems from Cocteau's tendency to take the final syllable of one line and use it to begin the next. This technique creates innumerable puns. In the first song, for example, the line 'Les clowns fleurissent du crotin *d'or*' ('The clowns flourish on golden dung') is followed by '*dor*mir' ('to sleep'). In the second song, the words 'piano méca*nique*' prompt *Nick* Carter, and 'Un bonjour Gust*ave*' precedes '*Ave* Maria de Gounod'. Since Gounod based his *Ave Maria* on J. S. Bach's Prelude in C Major from the *Well-Tempered Clavier*, Book 1, Poulenc takes the opportunity to insert a C major scale moving in quick sixteenths in the vocal line (see Ex. 6.13, bar 2). Occasionally Cocteau transforms sense into

Ex. 6.13. Allusions to Bach and Gounod in *Cocardes*

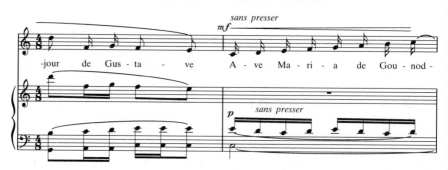

nonsense by rewriting words so that 'moi' ('me') becomes 'moa' (a neologism) and 'voulez' ('want') becomes 'volez' ('steal'). The utter lack of subtlety in Cocteau's manipulation of the text contributes to the hilarity and imparts an irreverence reminiscent both of Satie's

[33] Poulenc explained to Claude Rostand that his small band accompaniment 'responded exactly to the fairground style desired by Cocteau'. See Poulenc, *Entretiens*, p. 66. The original orchestration is available only on location at Max Eschig Publishers in Paris. The discussion in this chapter is based on the present author's study of the score at Eschig.

comical annotations in his piano pieces of the 1910s and of Dadaist word games.

Poulenc's orchestration in *Cocardes* gives the brass a prominent role. The cornet plays the melody, the trombone is the bass instrument, and the violin part contains ostinato figures. In the third song, Poulenc's brass writing creates imaginative jazz colourings. The trombone opens with chromatic glissando lines and later embellishes a sustained bass line with three grace notes that ascend chromatically (see Ex. 6.14a–b). The resultant slide imitates a

Ex. 6.14. Jazz colouring in the brass writing of *Cocardes*

(a) Chromatic glissando lines

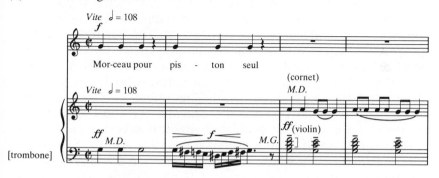

(b) Chromatic grace notes

characteristic New Orleans trombone style that Poulenc probably discovered through contact with the Hellfighters and Mitchell's Jazz Kings, or with Parisian bands that absorbed the technique.

Despite Poulenc's authentic use of brassy orchestration to imitate a small dance-hall band, his musical language in *Cocardes* seems to be at odds with the style of popular song. Whereas popular tunes generally follow a conventional pattern of stanzas or verse and refrain and observe a regular metre and a tuneful style, the songs in *Cocardes* are through-composed with frequent and sudden shifts in tempo and occasional excursions into vocal declamation. Poulenc's source for these seemingly unconventional devices stemmed once again from the performance style of Maurice Chevalier. In the opening of the first song, 'Miel de Narbonne', for instance, several measures of abrupt changes in tempo and metre precede the establishment of a regular pulse, and tempos continue to shift, sometimes in rapid succession, throughout. The third song, 'Enfant de troupe', begins with similar tempo and metre fluctuations and proceeds to re-create Chevalier's declamatory style by applying a rhythmic pattern of successive eighth notes to pitches lying a step apart (see Ex. 6.15, bars 1–2). The effect suggests not only Chevalier but recitative since the declamatory vocal line appears above an accompaniment in which rhythmic flow is suspended.

A cheerful spirit and a melancholy tone are two seemingly contradictory aspects which pervade the songs in *Cocardes*. The two can be reconciled as expressions of Poulenc's nostalgia, a tender sentiment associated with memories of boyhood experiences in music-halls and *bals-musettes*. Central to the notion of nostalgia is the contrast between one's present situation and sentimental memories of the past, between the surface gaiety of Parisian street life and the wistfulness beneath. A remarkable example of

Ex. 6.15. Elastic tempo and declamatory vocal line in "Enfant de Troupe"

Poulenc's capacity to express nostalgia musically and thus to leave the bittersweet impression observed by Paul Collaer occurs in the third song of *Cocardes*. Here the lively, boisterous music ('vite') that opens the piece returns in a middle section where its character and hence its mood are utterly transformed. The dynamic level has shifted from fortissimo to piano, the tempo is now slow, and a sorrowful violin marked 'triste' sings the melody previously proclaimed exuberantly by the trombone. Although Poulenc recalls the lively version of this music at the end, its sound is now coloured by the melancholy interlude.

Erik Satie experimented more frequently with musical parody than with nostalgia. The 'spoofing' of art music which enlivened *Parade*, *Sports et divertissements*, and many of the piano pieces of the 1910s is taken up with particular relish in *Trois Petites Pièces montées*, appearing next on the programme of the 'Spectacle-Concert'.[34] Satie's penchant for parody, especially one aimed at academicism, reflects the influence of cabaret humour on his approach to musical composition and parallels the Dada movement.

The conclusion to the third piece of *Trois Petites Pièces* offers one of Satie's most undisguised concert music 'spoofs'. First, he slows down the pace considerably by sustaining the pitch A, as shown in Ex. 6.16a–b, bars 1–4. Then he presents a dialogue in which the two piano parts exchange unduly simple, two- and three-note ideas. The humour of the passage stems from Satie's repetition and embellishment of these banal musical figures and from his application near the end of swinging dotted rhythms in the bass in imitation of early jazz (Ex. 6.16a–b, bars 18–21). Satire is reinforced, moreover, because Satie follows his dialogue with the characteristic strains of nineteenth-century music: chromaticism in the bass-line and reiterated cadential chords (bars 22–7) which bring the piece to an emphatic close. An important precedent for this mockery of eighteenth- and nineteenth-century protracted cadences appears in Satie's *Embryons desséchées* (1913) in which a 'compulsory cadence [the Author's]' occupies the entire last page of the work. Stravinsky engaged in a similar parody when he concluded *Pulcinella* (1919) with a prolonged repetition of one cadential pattern.

The final work on the programme of the 'Spectacle-Concert', Milhaud's *Le Bœuf sur le toit*, rounded out an evening as diverse as those spent in Parisian music-halls. The audience had witnessed a

---

[34] The *Trois Petites Pièces montées* were first written for small orchestra (flute, oboe, clarinet, bassoon, horn, two trumpets, trombone, strings, percussion without drums) and were performed in that version at the 'Spectacle-Concert'. Examples of this piece will be taken from Satie's four-hand piano arrangement.

Ex. 6.16. Parody in *Trois petites pièces montées*

(a) Prima

(b) Seconda

foxtrot, a round of songs or *tour de chants*, and a satiric instrumental piece. Now they were entertained with a work that resembled a music-hall revue in its use of costumes, scenery, visual spectacle, and orchestral accompaniment. Milhaud composed *Le Bœuf* in 1919 as a fantasia intended to accompany one of Charlie Chaplin's silent films. When Cocteau heard the music, he resolved to use it instead as the accompaniment for a stylized ballet-pantomime.[35] Accordingly, he imagined a scenario set in an American bar during Prohibition and featuring a variety of eccentrics—a Lady of Fashion, a Red-headed Woman dressed as a man, a Barman, a Policeman, a Negro Boxer, a Jockey, a Gentleman in evening clothes, a Negro Billiards Player—all of whom amused themselves by sipping cocktails and dancing. Since Milhaud did not write his fantasia with Cocteau's scenario in mind, there was no correspondence between action and music. Yet Cocteau embraced the lack of congruity and accentuated it by choreographing very slow movements in contrast to the lively tempo of the music.[36]

Le Bœuf sur le toit took its title from a Brazilian popular melody. The tunes Milhaud assembled for his ballet-pantomime were based on Brazilian songs, tangos, maxixes, and sambas that he had come to know during his stay in Rio de Janeiro (1917 to 1919) and that he transcribed upon his return. Milhaud presents these melodies in pairs between recurring statements of a Brazilian rondo theme. He consistently applies duple metre and a sectional design, both common traits of popular dances. A nine-bar section bearing the rondo theme is followed by a seven-bar strain of transitional music. Then, after two sixteen-bar Brazilian dance or song tunes, the rondo theme returns and the pattern repeats itself. The piece thus assumes a large-scale formal pattern inspired by the design and phrasing of popular dance music.

Milhaud's familiarity with Brazilian dance styles prompted his use of rhythmic mannerisms which evoked specific Brazilian dances. The rondo theme, illustrated in Ex. 6.17, and other tunes in the piece contain the characteristic sixteenth–eighth–sixteenth rhythm of samba melodies. Milhaud also suggests the Brazilian tango—rhythmically equivalent to the Cuban habanera—by employing dotted and syncopated figures in his accompaniments. Syncopation is based once again on the rhythm of sixteenth–eighth–sixteenth. Milhaud may have studied tango rhythms in Rio or in Paris where

---

[35] Milhaud describes the genesis of *Le Bœuf* in *Notes without Music*, pp. 86–7.
[36] Cocteau's conception of a lack of correspondence between music and scenario prefigures the experimental dance pieces of Merce Cunningham and John Cage.

Ex. 6.17. Samba rhythms in the rondo theme of *Le Bœuf sur le toit*

the tango style (with urban Argentinian sources) grew popular around the turn of the century.[37]

The steady percussion accompaniment in *Le Bœuf* is provided by an instrument which Milhaud labelled 'guitcharo' in the score. Described in *Notes without Music* as a 'long gourd on which a series of grooves have been traced', the 'guitcharo' is most likely Milhaud's remembered term for the *guiro*. This choice of instrumentation, like the dance rhythms, reveals Milhaud's close adherence to Brazilian models.

Le Bœuf sur le toit also contains abstract allusions to the popular world and, more generally, to the sounds of modern urban technological life. Milhaud was dazzled by the simultaneity of the Parisian fairground and the Brazilian Carnaval, where a complex network of patterns and sights overwhelmed his senses. Throughout his ballet, he employs polytonality as a musical metaphor for this jangling simultaneity, for the chance combinations of sounds we hear when two or more steam organs are playing simultaneously in different keys. H. H. Stuckenschmidt observes that in *Petrushka*, Stravinsky used a similar device of joining and juxtaposing different tonalities in imitation of the fairground medley.[38] In *Le Bœuf*, however, the technique of bitonality is expanded for one sometimes hears three and even four tonalities at once. Ex. 6.18 illustrates a passage in which samba-style flute melodies in F major sound against a horn and oboe melody in F minor and a violin melody in the distant key of A major. The melodies that Milhaud superimposes employ simple tonic and dominant harmonies; when combined, however, their sound is highly dissonant. The same mixture of sounds, each simple in itself yet clamorous in combina-

---

[37] The urban Argentinian tango was equivalent to the Cuban habanera which Brazilians named 'tango' in the latter part of the nineteenth century.

[38] See H. H. Stuckenschmidt, *Twentieth-Century Music*, trans. Richard Deveson (New York: McGraw-Hill, 1969), p. 79.

Ex. 6.18. Polytonality in *Le Bœuf sur le toit*

tion, pervaded the Parisian fairground and the streets of modern Paris.[39]

Milhaud's choice of a polytonal language was radical in its aural and visual depiction of the modern urban world. Polytonality also offered an alternative to German chromaticism and allowed him to explore a different kind of dissonant language, while retaining the diatonic, folk, and popular materials that affirmed his ties with popular entertainment.[40]

With the last strain of Milhaud's rondo theme, the 'Spectacle-Concert' came to a close. According to Milhaud, critics and public alike reacted to the performance by labelling him, in his words, a 'writer of comic circus music'. Milhaud was dismayed by this reaction for, if he had written *Le Bœuf* as a 'merry divertissement',

[39] Certainly *Le Bœuf* did not represent Milhaud's first experiments with polytonality, which date back to the composition of *Les Choéphores* in 1915. The imitation of a fairground arena is particularly strong in *Le Bœuf*, however, because the individual melodic strands which Milhaud superimposes feature simple tonic and dominant harmonies and samba rhythms.

[40] H. H. Stuckenschmidt quotes Ernst Krenek on the possibilities that polytonality offered Milhaud in terms of both modernist dissonance and a style based on folk song and neo-classicism. See Stuckenschmidt, *Twentieth-Century Music*, p. 79.

he had certainly not intended it to be laughable. Eight years after the 'Spectacle-Concert', Milhaud's denials of frivolity were reinforced by the critic Henri Prunières who gave serious acknowledgement both to the music-hall event and to the aesthetic which it represented:[41] 'This memorable presentation marked the triumph of Cocteau's ideas. The circus, jazz, the cinema, and finally and above all the music-hall, furnished the principal elements of the new aesthetic, which enjoyed but a short vogue.'

WEAPON AGAINST THE 'FALSE-SUBLIME': *LES MARIÉS DE LA TOUR EIFFEL* (1921)

The 'new aesthetic' continued for several years after the 'Spectacle-Concert'. In 1921 Cocteau created his *pièce-ballet*, *Les Mariés de la tour Eiffel*, which contained dialogue in place of the scenarios that accompanied *Parade* and *Le Bœuf sur le toit*. This dialogue, and indeed the entire conception of *Les Mariés*, was the product of Cocteau's most far-reaching exploration of everyday themes and his most sweeping blend of diverse forms of entertainment. The absurd elements in the story line and the satiric humour of the music served once again as weapons against the current 'preoccupation with the false-sublime'.[42]

Cocteau created *Les Mariés* in response to a commission from Rolf de Maré, the director of the Ballets Suédois, who requested a *pièce-ballet* (play-ballet) similar to the eighteenth-century French *opéra-ballet*. De Maré asked members of the 'Six' to contribute musical numbers, and all but Louis Durey complied. The choreography, consisting of the same pantomime movement featured in *Parade* and *Le Bœuf*, was devised by Cocteau and Jean Borlin. Irene Lagut designed the décor, Jean Hugo the costumes, and on 18 June 1921 the work received its première at the Théâtre des Champs-Élysées.

Cocteau realized de Maré's vision of a *pièce-ballet* by creating a contemporary analogue to the blend of spectacle, ballet, music, chorus, and song that constituted the *opéra-ballet*. Cocteau's own words best describe the new form he invented: 'a kind of secret marriage between ancient tragedy and the *revue de fin d'année* (end-of-the-year revue), chorus and music-hall number'.[43] Both the *revue*

[41] Henri Prunières, 'Francis Poulenc', *The Sackbut* 8 (1928): 192.

[42] This quote is taken from Cocteau's 'Preface' to *Les Mariés de la tour Eiffel* which appeared in 1922. For a translation by Dudley Fitts, see Jean Cocteau, *The Infernal Machine and Other Plays* (New York: New Directions, 1963), pp. 153–60.

[43] This description appeared in the ninth issue of the revue *La Danse* and is quoted in Jacques Pradère's liner notes to *Les Mariés de la tour Eiffel*, Ades 14.007. Translation is by the present author.

de fin d'année—which had its origins at the Parisian fair in the eighteenth century—and the chorus of ancient tragedy are represented in Cocteau's use of two human phonographs that assume the roles of *commère* and *compère* and narrate the events. Cocteau expands their roles so that they also speak for the different characters in the play. From the music-hall, Cocteau borrows the device of living postcards or *tableaux vivants* such as those featured in the *revue à grand spectacle*. Hence the Trouville Bathing Beauty of *Les Mariés* emerges from the camera with coloured lights and is admired by the wedding party as a 'pretty postcard'. Music-hall elements in *Les Mariés* also take the form of dances, ranging from a 'Waltz of the Telegrams' to a 'Wedding March' and a 'Dance of the Trouville Bathing Beauty'. Fairground games are transferred, moreover, in the comical scene in which the child of the bride and groom tosses bullets and 'massacres the wedding party'. The child's actions imitate the routines of a *jeu de massacre* player at the Parisian fair.[44]

In the preface written for the published play of *Les Mariés* in 1922, Cocteau demonstrated his kinship with Apollinaire, and specifically with the poet's call to 'unveil truths', when he explained that his aim was to 'disengage' and 'rejuvenate the commonplace'. Juxtaposed in *Les Mariés* against the quite ordinary circumstance of a bourgeois wedding party that wanders to the lower platform restaurant of the Eiffel Tower to have its picture taken, is the extraordinary situation of phonographs that talk and a camera that has a will of its own and ejects its subjects. Commonplace sayings like 'Watch the camera. Look at the birdie' and nonsensical ones such as the child's 'I wanna feed the Eiffel Tower' are taken quite literally: a live bird (an ostrich) actually pops out of the camera, and the child is informed that 'The Eiffel Tower is only fed at certain hours. That's why it's behind bars.' Cocteau ridicules clichés by heaping five or six hackneyed sayings in quick succession, with the culminating effect that they all contradict one other and say nothing. The child is thus praised as 'the image of his mother', of his father, of his grandpa, of his grandma and of everyone else in sight.[45] As part of the exploration of truth and surprise promoted by Apollinaire, Cocteau also accentuates the fine line between animate and inanimate, real and mirage. The radiograms turn into burlesque performers, while the wedding party becomes a painting with a 'sold' sign placed in front of it at the end.

---

[44] For information on the *jeu de massacre*, see Ch. 1.

[45] Since 'cliché' in French also means 'snapshot', the end to this string of clichés comes only when the photographer manages to take a picture of the wedding party.

The musical numbers composed by Auric, Milhaud, and Poulenc to accompany Cocteau's play employ popular styles and often use musical caricature to reinforce the formal pretensions, the sentimental excess and the absurd behaviour which characterize the wedding party. An 'Ouverture' by Auric sets the mood. Thuds in the tuba, timpani, and double bass punctuate a heaving march which blatantly debunks the grace we expect of a wedding procession. In a forecast of the music-hall and fairground elements to come, Auric then weaves countermelodies in different keys around a sturdy brass fanfare in order to convey the din of the fairground. The polytonal medley disperses and gives way to a series of charming folk tunes similar to those a stroller might hear when listening to individual themes at the fair.

Milhaud's 'Marche nuptiale' also features a brass fanfare and juxtaposes different tonalities, but achieves a sharper clash of sounds than Auric's piece because three different tonal layers are sometimes heard at once. The march accompanies the entrance of the wedding party. As each member arrives, the narrators introduce them with a cliché: 'The bride, sweet as a lamb'; 'The bridegroom, handsome as a matinée idol'; 'The General, dumb as a goose'. Milhaud creates musical counterparts for these hackneyed expressions in the form of a 'syrupy' violin solo that follows wide leaps and ascends to the peak of the violin range (see Ex. 6.19, bars 1–3).

Ex. 6.19. Satire of clichéd expressions in "Marche nuptiale"

Another musical satire, this time of ceremonial pomp, appears at the end of Milhaud's 'Marche', where a pattern of C major chords is reiterated fortissimo in strings and brass until the piece comes to a melodramatic close. Milhaud's extension of the cadence mocks not only ceremonial pomp but the eighteenth- and nineteenth-century techniques which Satie and Stravinsky likewise satirized in the prolonged cadences of *Embryons desséchés*, *Trois Petites Pièces montées*, and *Pulcinella*.

Caricatures of pomp are also the subject of Poulenc's 'Discours du Général' ('General's Speech').[46] According to Cocteau's stage

---

[46] Poulenc reorchestrated and revised both of his contributions, 'Discours du Général' and 'La Baigneuse de Trouville', in 1957, and this is the only version available in score. It is in Poulenc's hand. A study of the original 1921 manuscript, on microfilm at the Bibliothèque

directions, this number is to be played in lieu of a verbal speech. While the music proceeds, the General stands before the wedding party and gesticulates, just as a character in a silent film communicates a message with gestures and facial expressions. Poulenc conveys the General's bravado by punctuating the principal cornet theme with snare drum rolls and presenting responses to the cornet lines in the form of swaggering trombone glissandos (see Ex. 6.20). In addition he mocks the General's pretensions to dignity by employing polka rhythms and a lively polka tempo in place of the march style appropriate for a General. The vacuousness of the

Ex. 6.20. Satire of pomp in "Discours du Général"

Nationale in Paris, however, reveals the principal differences between the two versions: the 1921 version uses clarinet instead of oboe in 'Discours du Général', and is brassier and sometimes more contrapuntal in 'La Baigneuse'. Examples of this piece will be taken from the 1957 score. Examples of Milhaud's *Le Train bleu* (discussion to follow) will be taken from the 1924 manuscript in Milhaud's hand, published by Heugel et Cie in Paris.

speech is conveyed in the cornet theme itself which repeats each pitch once and sometimes twice before moving on to the next (see Ex. 6.20). After hearing such banal music, the phonographs announce that 'everyone is deeply moved'—a cliché which sarcastically accentuates the speech's platitudes.

'La Baigneuse de Trouville', Poulenc's next musical number, is a frenzied dance in duple metre with a *valse lente* interlude. This music follows the surprise appearance of a Trouville Bathing Beauty—rather than a little bird—from the Photographer's camera.[47] To the wedding party's delight, she poses as a music-hall *tableau vivant*, made radiant by coloured lights and bearing a butterfly net with a heart and a picnic basket. For the accompaniment to her dance, Poulenc incorporates popular styles comprising a four-bar introduction and vamp, an emphasis on brass and woodwinds which is even stronger in the 1921 orchestration than in the revision of 1957, and a choice of *valse lente*. He caricatures sentimentality by interrupting the lilting flow of the waltz with chromatic ascents in the woodwinds that induce excessive rubato.

THREE 'MONUMENTS TO FRIVOLITY' AND A BLUES BALLET

Traditionally the collaboration on *Les Mariés de la tour Eiffel* has been said to mark the end of the circus, jazz, and music-hall aesthetic identified by Henri Prunières.[48] Yet an important repertoire of ballets composed after 1921 reveals that Satie, Milhaud, Poulenc, and Auric continued to embrace popular genres and to challenge Impressionism through the year 1924. At the same time, certain important changes took place. While the early songs and instrumental pieces of the younger composers drew directly from Erik Satie in their imitation of children's tunes, their repetitive quality, and their recourse to dance idioms, the ballets of 1921–4 showed a growing independence. Poulenc began experimenting with eighteenth-century styles and French operetta; Milhaud explored both operetta and early jazz; Auric substituted French folk and operetta tunes for his previous use of American syncopated dance idioms and blues. If the basic tenet of a French music freed from nineteenth-century German and Impressionist influences still held, the models for creating tuneful music were now different and decidedly less American. In *Les Biches* (The Does), *Le Train bleu*,

---

[47] Trouville was one of the most popular seaside resorts in Normandy.
[48] Keith Daniel describes the dissolution of the group after 1921. See Daniel, *Francis Poulenc*, pp. 21–2.

and *Les Fâcheux* (The Bores), moreover, references to eighteenth-century conventions of periodicity, homophonic texture, and scalar writing reflected the emerging neoclassicism in Europe.

Les Biches (1924)

Serge Diaghilev attended the June 1921 première of *Les Mariés de la tour Eiffel*, and his interest in the contributions of Poulenc, Milhaud, and Auric prompted him to commission ballets from these three composers. Poulenc was the first to be approached. Through Igor Stravinsky, Diaghilev secured a meeting with the young composer and requested a ballet for the 1922–3 Monte Carlo season of the Ballets-Russes. Poulenc responded with his first ballet, *Les Biches*, which received its première on 6 January 1924, featuring sets and costumes by Marie Laurencin and choreography by Bronislava Nijinska.

Poulenc conceived of *Les Biches* as a modern-day *fêtes galantes*.[49] Rather than following the pattern of Diaghilev's earlier ballets and introducing a detailed scenario, he sought simply to create an atmosphere of charm and eroticism reminiscent of the mood conveyed in Jean-Antoine Watteau's paintings of eighteenth-century pastoral gatherings. Hence he avoided plot, specifying only the setting of the ballet—a large country drawing room with a huge sofa—and the characters who filled the room—twenty women and three handsome men. The characters' flirtations offered the sole suggestion of a story line. Darius Milhaud applauded Poulenc's radical approach: 'The novelty of this ballet is to have no scenario, no subject, nothing to describe, to suggest, to express, to comment upon. Absolutely no literary plot, no precise date for the costumes. We are in full fantasy.'[50] Implicit here was praise for the scenario's timelessness and for Poulenc's challenge of the close correspondence between music and ballet scenario in Benois's *Petrushka* and in other works of the Ballets-Russes. 'Fantasy' was a term that Milhaud had also applied to 'the extraordinary mixture of different ingredients' in *Les Mariés de la tour Eiffel*.[51]

In a statement made during conversations with Claude Rostand, Poulenc affirmed the importance of 'pleasure' and hedonism as governing principles in *Les Biches*. Two musical devices impart the gaiety and the wanton spirit that Poulenc desired: first, abrupt shifts in musical style, resulting in a heterogeneous pool of idioms;

[49] Poulenc explains this conception in Francis Poulenc and Stéphane Audel, *My Friends and Myself*, trans. James Harding (London: Dennis Dobson, 1978), p. 43.
[50] See Milhaud's *Études*, p. 64.
[51] See *Notes without Music*, p. 95.

second, a tunefulness derived from eighteenth-century classicism, early nineteenth-century Italian operatic styles and late nineteenth-century French operetta (Offenbach and Messager).[52] The stylistic diversity indicates Poulenc's ability to transfer the many sounds of popular and concert music that he encountered during his youth, as well as the diversity and simultaneity inherent in Parisian popular entertainment, to an art composition. The specific choice of styles represents, in part, Poulenc's response to Diaghilev's interest in eighteenth-century ballets (Tommasini's *The Good-Humoured Ladies* and Stravinsky's *Pulcinella*) and in ballets based on such early nineteenth-century composers as Rossini. Like *Les Mariés de la tour Eiffel*, moreover, *Les Biches* combines dance, music, song, and spectacle in a modern-day version of the eighteenth-century *opéra-ballet*.

From the outset, the procedure of contrast serves to propel the music of *Les Biches* forward.[53] Ex. 6.21 illustrates instances of stylistic juxtaposition that appear in the ballet's 'Ouverture'. Poulenc opens with a slow introduction that imitates the Russian manner of Stravinsky's *Les Noces* by using modal writing, grace notes suggestive of a glottal singing style, monophonic texture, and changing metres (see Ex. 6.21a).[54] In concluding, the section duplicates the procedure of Haydn's slow introductions and rests on the dominant chord before proceeding with a sprightly 'Allegro molto'. The Allegro ushers in an entirely different musical style, suggestive both of eighteenth-century classicism and the early nineteenth-century operatic language of Rossini. Aspects of classicism manifest themselves in the presence of antecedent and consequent units (Ex. 6.21b), the application of sequence (Ex. 6.21c), and the occasional trace of a linear Alberti bass accompaniment. Rossini's language can be heard in the continuous stream of melody, the recurrent use of one or two rhythmic motifs, the brevity and regularity of the

[52] As late as 1944, Poulenc's passion for turn-of-the-century French operetta prompted him to lift half-a-dozen bars of the number 'Ah que ne parliez-vous?' from André Messager's *La Basoche* and insert them unchanged into his *opéra bouffe*, *Les Mamelles de Tirésias*. I am indebted to Roger Nichols for calling my attention to this information which appears in James Harding, *Folies de Paris* (London: Chappell, 1979), p. 127.

[53] The following analytic discussion focuses on the complete ballet rather than on the orchestral suite. Poulenc revised and reorchestrated *Les Biches* in 1939–40. A piano score of the original 1923 version, available on location at Heugel in Paris, reveals few discrepancies in pitch between 1923 and 1939 and indicates that the blue notes and syncopation of the 'Adagietto' were choices made at the time of composition. It is impossible to study the original orchestration, however, because all but nine measures of the manuscript were destroyed. Since the 1939 revision represents changes in orchestration made after the period under discussion here, matters of orchestration in *Les Biches* will not be addressed.

[54] The grace notes may refer to eighteenth-century manners as well as to Stravinsky's *Les Noces*.

Ex. 6.21. Stylistic juxtaposition in the "Ouverture" to *Les Biches*, Poulenc's piano reduction

(a)

phrases which are frequently repeated, and the clear simple harmonies.[55]

The elegant, witty mood of Mozart and Rossini is interrupted by Poulenc's insertion of a waltz more striking for its individual stamp than its adherence to traditional waltz styles. Poulenc sets his waltz in 6/8 metre, instead of the graceful 3/4, so that it moves at about twice the customary speed (see Ex. 6.22). He replaces the characteristic triple time (oom-pah-pah) accompaniment—in which each beat of the bar is defined by a separate punctuation, first note, then chord–chord—with a fortissimo chordal articulation on the first beat only. Yet the alternation between this tonic chord and the dominant chord that articulates the second three-beat grouping in each bar imitates the style of popular accompaniments. In addition, the pitch repetitions in Poulenc's melody and the lilting style exude a light, popular mood true to the waltz and different from the styles heard thus far. Poulenc continues with a return to the preceding classical-inspired themes and then, in conclusion, a recall of the 'Ouverture's slow introduction.

Ex. 6.22. The waltz in the "Ouverture" to *Les Biches*, Poulenc's piano reduction

The diverse stylistic range in the 'Ouverture' offers a preview of the succession of different idioms to come—operetta in the 'Rondeau', blues and American syncopated dance music in the 'Adagietto', foxtrots in the 'Rag-Mazurka', nineteenth-century Russian chorus in 'Jeu', quotations of Mozart in the 'Finale'. Indeed Poulenc's method of using the overture to encapsulate the ballet's

---

[55] For an example from Rossini's œuvre that offers a stylistic model for this opening portion of *Les Biches*, see the overture to *L'Italiana in Algèri*.

diversity resembles the common operatic device of treating the overture as a potpourri of prominent themes in the opera. Yet with *Les Biches*, the emphasis is on a variety of styles rather than themes. Tunes presented in the overture, moreover, do not return again until the finale.

The success of *Les Biches* stems from Poulenc's ability to incorporate a vast array of idioms and at the same time avoid diffuseness. Poulenc achieves a unified ballet by emphasizing the common elements in some of the styles he juxtaposes. For instance, eighteenth-century classicism and early overtures of Rossini share with Offenbach's and Messager's music a reliance on simple folk and dance tunes that brings with it periodic phrasing, scalar and triadic writing in the melody, rhythmic drive, and firm tonal orientation.[56] Certainly the complexity of Mozart and Haydn arises from the manipulation of periodicity and the use of chromaticism and counterpoint, but Poulenc emphasizes the predictable features of classical music. Thus there is a fine line in *Les Biches* between a classically oriented theme and one that suggests popular sources.

In passages which contain a closer affinity to popular than to classical music, moreover, the specific popular source is often difficult to pinpoint because Poulenc tends to diverge subtly from the regularity of an operetta or music-hall tune. The principal theme of the 'Rondeau' movement, shown in Ex. 6.23, is a case in point.

Ex. 6.23. The principal theme of the "Rondeau", Poulenc's piano reduction

[56] The identification of folk music with symmetry of rhythm and phrasing is not always true, of course, but the symmetrical aspects of folk music are the traits emphasized by Haydn when he employs a folk style, and by Offenbach.

The tune employs short durations, follows a rising scalar pattern and major mode typical of Offenbach melodies, and appears within a homophonic texture. Successive repetition of pitches coupled with the antecedent-consequent structure creates a simple, memorable melody. However, the accentuation of the fourth beats in bars 2 and 4, followed by syncopation, disturbs the metric regularity and transforms traditionally weak beats into downbeats. In this way the listener hears a new bar begin with the final beats of these two bars. Poulenc's barring in 4/4 thus disguises an ambiguity in the number of beats per bar. Such ambiguity would have been uncharacteristic both of late nineteenth-century operettas of Offenbach and Messager and of early twentieth-century operettas and revues of Christiné and Yvain.

The syncopation in the 'Rondeau' theme of *Les Biches*, although not typical of French operetta music, serves to enhance the theme's popular quality. Poulenc may have been inspired by his contact with American dance music. The next movement in the ballet, the 'Adagietto', makes more extensive use of syncopation. The influence of American music is confirmed by Poulenc's evocation of 'blue notes'. Ex. 6.24 illustrates an eight-bar passage from the 'Adagietto' in which the accompaniment consists of untied syncopation in the cellos and a quarter note underpinning in the double bass. A chromatic descent in the cellos combined with strong root movement in double basses produces a succession of dominant sevenths that follow a circle of fifths (F♯–B–E–A). Above, the bass clarinet presents a melody which suggests 'blue' notes both in its semitonal movement and, in bar 6, in the stepwise descent from D♯ to C♯ to B followed by a leap to D (bar 7, first beat) as flat three in the key of B major.[57] The fusion in these bars of a blues-style bass clarinet melody with a syncopated accompaniment featuring a simple dominant seventh chord progression offers a remarkable evocation of early jazz.

The Adagietto of *Les Biches* stands alone in Poulenc's œuvre in its allusions to early jazz. This exceptional position is consistent with Poulenc's impatient attacks on a style he felt was becoming far too fashionable in Paris. 'I don't even like jazz,' he wrote, 'and I certainly don't want to hear about its influence on contemporary music.'[58] Fittingly he crept it, not into the 'Rag-Mazurka' where we might have expected a syncopated dance idiom, but into a movement whose title betrays no jazz influence. This can be

[57] References to pitches in the bass clarinet are as transposed, not as written. The F in bar 6 is transposed as D♯ rather than E♭ because of the harmonic underpinning of a dominant seventh chord on B.

[58] Poulenc's attack on jazz is quoted in Keith Daniel, *Francis Poulenc*, p. 23.

Ex. 6.24. "Blue notes" and syncopation in the "Adagietto"

contrasted with the inconsequential use of dotted rhythms in the Rag-Mazurka. The models for Poulenc's incorporation of a blues style in the Adagietto probably encompassed recordings and live performances by Louis Mitchell's Jazz Kings, performances featuring Sidney Bechet with Will Marion Cooke's Southern Syncopated Orchestra, and concerts by Billy Arnold's Novelty Jazz Band.

In a statement that Cocteau wrote on *Les Biches* in 1924, he described the 'beauty and melancholy' of the piece and added, 'I doubt whether this music knows it hurts'.[59] The nostalgic component

[59] See Jean Cocteau, *The Cock and the Harlequin* in *A Call to Order*, trans. Rollo Myers (New York: Haskell House, 1974), p. 65. In his *Études*, p. 67, Milhaud comments similarly on the sadness behind the hearty façade, the grandeur which transforms into a veritable anguish.

of the ballet, the tragic sound lurking beneath the tuneful surface, is a feature sensed immediately by the listener and noted by commentators. Poulenc's technique of creating this tone is more subtly applied than in the song cycle *Cocardes* and consists of contrasting a loud, vibrant section with a few bars of solo wind playing in a high register. The wind portions often contain expressive markings of 'very sweet' or 'melancholy', as in the clarinet of Ex. 6.25, and sound especially mournful because they interrupt boisterous music-making—in this instance, the fully orchestrated, choral opening of the 'Chanson dansée'. The solo clarinet in Ex. 6.25 serves both to shift the mood and to provide a transition to a more sombre section of the song. Solo wind instruments here and elsewhere (rehearsal number 104, for example) shade *Les Biches* with a poignant, bittersweet hue.

Ex. 6.25. Nostalgia in *Les Biches*

Les Fâcheux (1923) and *Le Train bleu* (1924)

Darius Milhaud was intrigued by the fantasy which emerged when ballets had no scenarios, no specifications of plot or setting. He admired this quality and the accompanying hedonism of *Les Biches*, and explored both when he sat down in 1924 to write the music for a ballet commissioned by Diaghilev, *Le Train bleu*. Cocteau's scenario did not specify any particular action, but suggested an atmosphere of playful frolicking. Bathers and gigolos embarking from the elegant blue train arrive at a fashionable resort where they amuse themselves with tennis, golf, and other favourite sports. In excerpts from some notes on the scenario, Cocteau wrote, '*Le Train bleu* must not be a light work, but a monument of Frivolity!'[60] The paradoxical elevation of 'Frivolity' to monumental and, indeed, masterpiece status clearly delighted Cocteau. The scenario for Auric's *Les Fâcheux*, a Diaghilev commission of 1923, was equally simple. Adopted by Boris Kochno from Molière's play of the same

---

[60] Quoted by Paul Collaer in *Darius Milhaud* (Geneva: Éditions Slatkine, 1982), p. 128.

name, it concerns the escapades of Eraste who is detained by gossips, card-players, and other 'bores' on his way to an assignation.

With such inconsequential stories, the composers needed only to capture the appropriate mood. Accordingly, the music of both ballets pursues a path of unbroken lightness, without the intrusion of nostalgia. In diary notes written on the day he started *Le Train bleu* (which was composed in a whirlwind two weeks between 12 February and 5 March 1924), Milhaud explained, 'I would like a music of laziness, nonchalant, very "Monsieur who whistles softly under his breath while strolling with his hands in his pockets and while winking his eye", genre Yvain . . .'[61] The same playful indifference pervades *Les Fâcheux*. Musical themes in both ballets evoke operetta, French nursery and folk songs, and circus fanfares. Repetition is extensive and is approached differently by each composer. Milhaud tends to repeat two musical styles in alternation, in a formal design of alternating statements and episodes (ABABAB). Auric's formal scheme is less systematic, but involves literal recalls of entire sections.

Both composers explore the realm of fantasy so appealing to Milhaud by presenting a bewildering assortment of styles in their ballets. The mixture also invokes the seventeenth-century fantasia, an instrumental medley of diverse popular airs, and dance and operatic themes. In a twentieth-century version of this musical potpourri, *Les Fâcheux* jumps swiftly from imitations of folk dances and children's tunes, to march and operetta themes, echoes of Stravinsky's *Pulcinella*, and parodies of late nineteenth-century German sonorities. Auric also introduces a cinematic diversity reminiscent of Satie's *Parade* by presenting short themes that rush by and interrupt one another abruptly, much as popular tunes are fragmented and overlap when one walks the Paris streets. Milhaud avoids this sudden juxtaposition, but is equally eclectic in his choice of musical material. The Brazilian samba rhythm of the opening is followed by variants on operetta melodies, nursery songs, and a neoclassical fugue framed by a brassy fanfare at the end. Polytonal clanging turns this fanfare into a dissonant medley that re-creates the simultaneity of modern Paris.

Auric and Milhaud were intrigued by Poulenc's choice of French operetta as a popular source in *Les Biches*. The principal theme at the opening of *Les Fâcheux*, shown in Ex. 6.26*a*, reveals a kinship with French operetta and with Poulenc's Rondeau. Foreshadowed in fragmentary form in the six-note idea of the introduction, Auric's

---

[61] Maurice Yvain was a Parisian composer of such popular music-hall songs as the erotic 'Mon Homme', performed by Mistinguett.

Ex. 6.26. Allusions to French operetta in *Les Fâcheux* and *Le Train bleu*

(a) *Les Fâcheux* (opening theme)

(b) *Le Train bleu* (polka theme)

(c) *Le Train bleu* (waltz theme)

tune features the rising scalar pattern, the repeated pitches, the strict tempo, and the short durations of Offenbach's melodies. Auric departs from the symmetry of Offenbach, however, as Poulenc does, by stopping abruptly on the third pitch and avoiding periodic phrasing. Melodies in *Le Train bleu* present a more direct reference to operetta, without the disturbances in symmetry created by Poulenc and Auric. Ex. 6.26*b–c* illustrate the polka theme from movement five and the waltz from movement seven. The affinity to

French operetta tunes is enhanced by the pungent brass reinforcements of the waltz theme and by the trumpet scoring of the polka melody.

Milhaud's close adherence to operetta styles was intentional; his diary notes from 16 February describe *Le Train bleu* as an 'opérette dansée (sans chant) . . . music in the genre of Offenbach, Maurice Yvain and a finale Verdi . . . Not one syncopation. This is Paris, dirty trick, lewd and sentimental, then very much polka, galop, waltz, etc.'[62] The reference to Paris is significant for, aside from the Brazilian samba tune of the opening, Milhaud turned exclusively to French popular sources in *Le Train bleu*. Auric similarly re-created French folk tunes, circus fanfares and marches, but did not once incorporate American popular idioms.

La Création du monde (1923)

Auric's devotion to French sources was consistent with his statement in 1919 upon completing *Adieu, New York!* that it was time to renounce American syncopated dance music. For Milhaud, on the other hand, the turn to French operetta was an isolated instance. In fact, Milhaud argued in his *Notes without Music* that treating this particular 'frothy' scenario of Cocteau required music in the manner of Offenbach—not his usual style. Milhaud believed strongly in the establishment of a 'clearer, sturdier, more precise type of French art'[63] epitomized by Satie's *Parade*, and he discovered classical ideals of simplicity and restraint in American and Parisian popular models alike. Both offered a powerful antidote to the presumed sublimity of German and Impressionist music. Yet of the four composers, Milhaud tended to favour popular idioms of foreign cultures—Brazilian and Black American—because they were exotic. The idea that 'Negro' jazz originated in the 'savage rhythms' of African music had tremendous appeal. The two years preceding *Le Train bleu* saw Milhaud's composition of two works inspired by American syncopated dance music and blues: *Trois Rag-Caprices* for piano (1922) and the ballet *La Création du monde* (1923).

When it was completed in 1923, *La Création du monde* represented the most faithful and comprehensive reproduction of American popular styles yet attempted by a European composer. Milhaud set to work on the ballet as soon as he returned to Paris from his 1922 American trip, where he had attended performances by female blues singers and become acquainted with black musical

---

[62] Quoted by Collaer in *Darius Milhaud*, p. 125.
[63] See *Notes without Music*, p. 80.

theatre in Harlem. He was impressed by the orchestras performing in Noble Sissle's and Eubie Blake's *Shuffle Along* and in Maceo Pinkard's *Liza*. The instrumentation of *Liza* provided the model for *La Création du monde*. Milhaud duplicated Pinkard's scoring for flute, clarinet, trumpet, trombone, piano, percussion, string quartet with the viola replaced by a tenor saxophone, and double bass. His only change was the addition of oboe, bassoon, and horn.[64]

Unlike the trilogy of ballets composed by Poulenc, Auric, and Milhaud for Diaghilev's Ballets-Russes (*Les Biches*, *Les Fâcheux*, and *Le Train bleu*), Milhaud's *La Création du monde* was written to accompany a detailed scenario by Blaise Cendrars. The collaboration was an active one; upon Milhaud's return to Paris, Cendrars and the painter Fernand Léger whisked him off to the local *bals-musettes* where they discussed the African subject-matter of the new ballet, commissioned by Rolf de Maré for his Ballets Suédois. Cendrars had travelled to Africa and South America and in 1921 had published a 'Negro' anthology in which he translated a book of 'Negro' songs and poems. A serious and enthusiastic proponent of Africana and jazz, he intended to write a scenario narrating the creation of the world according to African folklore. It was exactly the kind of exotic subject that appealed to Milhaud, and it fitted well with the jazz score he wished to write based on the music he had heard in the United States. His memory of black theatre and club bands in Harlem was reinforced, moreover, by the bands of accordion, clarinet, cornet, trombone, and sometimes violin which played every night at the Parisian *bals-musettes*.

Cendrars's scenario contained five sections for which Milhaud composed five movements. The correspondence between action and music is by no means exact. Milhaud chose to preface his five movements with an introduction, for which there is no corresponding part in the scenario. The introductory music, performed before the curtain rises, features a haunting song in the saxophone that conjures up the scene of Cendrars's first section in which an inchoate mass of bodies encircled by three giant gods—Nzame, Medere, N'kva—awaits the creation of the world. When the curtain rises and Milhaud's first movement begins, the lively, blues-style fugue does not illustrate Cendrars's first section depicting the creation, nor is there a connection between Milhaud's soulful second movement and the creation of plants and animals that occurs in section two. Upbeat, syncopated dance music in the third movement, however, matches the round dance of Cendrars's third section depicting the new incantations of the African gods and the

---

[64] Milhaud's discussion of *Liza* appears in *Études*, p. 57.

birth of man and woman. Similarly the growing frenzy of Cendrars's dance of desire in the fourth section is accompanied by a progressively agitated dance played by full jazz orchestra with shrill woodwinds. The final movement returns to the blues of Milhaud's first and second movements, with no particular evocation of the lovers' kiss and the springtime described in the scenario.

La Création du monde presents an interesting compromise between Milhaud's ballets *Le Bœuf sur le toit* and *Le Train bleu*—which set a mood appropriate to the scenario but made no attempt at musical and dramatic development—and a new kind of ballet in which the music matched the exotic African story and moved forward in tandem with the action: new events tended to be accompanied by new sounds. Legér enhanced the evocation of exotic, primeval worlds by designing highly stylized costumes that contained either frontal and symmetrical, or profile views. He also emphasized simple, geometric, and linear forms and based many of his designs on giant versions of machine parts, thereby alluding to the power of the new technology—a contemporary form of exoticism. [See Plates 10–11.]

The music for *La Création du monde* is both effective, dramatic ballet music and an independent concert piece to be marvelled at for its learned use of the jazz idiom. The French jazz critic André Hodeir finds Milhaud's blues borrowings the most authentic aspect of his early jazz facsimile. In passages of *La Création*, Hodeir observes, Milhaud 'got close . . . to the true significance of the blue note—its instability'.[65] This instability is captured in the attraction between major third and blue note, typically performed as a slide. The subject of the Fugue in the ballet's first movement, shown in bars 1–6 of the double bass in Ex. 6.27, contains alternations between F♮ (blue note) and F♯ (major third) which lend themselves to such slide performances. Similarly the countersubject in E major presents the lowered seventh in a descending slide and the lowered third in the context of a chromatic ascent, as shown in bars 9–11.

Ex. 6.27. Sliding character of "blue notes" in the Fugue of *La Création du monde*

[65] See André Hodeir, *Jazz: its Evolution and Essence*, trans. David Noakes (New York: Grove Press, 1956), p. 253.

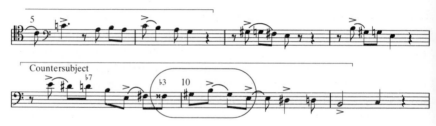

Within his countersubject, Milhaud weaves a five-note idea (circled in Ex. 6.27) which duplicates the opening figure of *St. Louis Blues*. Milhaud's familiarity with this famous jazz theme can be traced to performances by Jean Wiéner and Vance Lowry at the Bar Gaya. He probably also had an opportunity to hear recordings of *St. Louis Blues* on the Pathé, OKeh, and Columbia labels when he was in New York in 1922.[66]

The treatment of blues in *La Création du monde* demonstrates that Milhaud studied blues harmony as well as melody and understood an important element of a blues progression: the movement between tonic and subdominant chords. Milhaud employs this progression in a section of the second movement shown in Ex. 6.28, in which the lowered third (E♭) in the melody is heard above a subdominant chord (F major) that resolves to the tonic (C major). Both the subdominant and the tonic chords are coloured by blue notes on the seventh degree of the chord (with the pitch E♭ functioning as a lowered third in C major and a lowered seventh of an F major chord).

Rhythmically Milhaud adheres to early jazz styles in his consistent use of a duple metre (2/2). In addition, the third movement dance features Milhaud's remarkably authentic use of a device that ragtime composers called 'secondary ragtime'. This rhythm is created by superimposing a new rhythmic pattern upon the prevailing rhythm of the music. The pattern typically comprises a unit of three eighth notes placed above the normal four-quarter rhythm of jazz. Typically, also, the pattern follows an ascending or descending line and is repeated four times. Milhaud adheres to the secondary ragtime rhythm verbatim: he selects a unit of three eighth

[66] Hodeir mentions Milhaud's use of the opening figure of *St. Louis Blues*. Jim Europe's Hellfighters recorded *St. Louis Blues* on Pathé in March *1919* in New York. A group called Ciro's Club Coon Orchestra, featuring Vance Lowry on banjo, recorded the tune on Columbia in September *1917* in London. Lowry probably went on to introduce *St. Louis Blues* at the Bar Gaya. For this information on recordings of *St. Louis Blues*, see Brian Rust, *Jazz Records 1897–1942*, rev. 5th edn., 2 vols. (Chigwell, Essex: Storyville Publications & Co., 1982), i. 136, 315, 512.

Ex. 6.28. Blues harmonic progression in *La Création du monde*

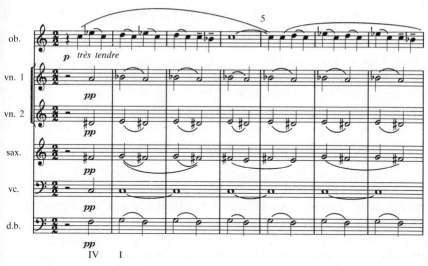

notes that moves in a descending line and suggests 3/8 metre, and he sets it above the prevailing four quarter note rhythm (see Ex. 6.29), repeating it five rather than four times. As Edward Berlin explains in his discussion of secondary ragtime, the result is not syncopation but rather a constant shifting of the accent within the superimposed three-note unit.[67] Milhaud's use of such ragtime rhythms may have stemmed from contact with the music of *Shuffle Along* and from his earlier encounters with Louis Mitchell's band in Paris.

Ex. 6.29. Ragtime rhythms in *La Création du monde*

---

[67] See Berlin's discussion of secondary ragtime in *Ragtime: A Musical and Cultural History*, pp. 130–1.

'NO PERFORMANCE': SATIE'S LAST TWO BALLETS

With each piece that he composed under the influence of American popular idioms, Milhaud selected a different style or genre as the basis for his popular re-creation. In *Caramel mou* he wrote a dance piece scored for jazz band and featuring the dotted rhythms of American shimmies and foxtrots. In *La Création du monde*, he employed the instrumentation of black theatre orchestras and the blues of Harlem singers recording on OKeh, Black Swan, and Columbia. In each case, Milhaud's choice of genre reflected the American music he had most recently encountered.

Erik Satie, by contrast, consistently applied stylistic features of the first American musical forms that had appeared in Paris in the twentieth century and the first that he became acquainted with: march and cakewalk. In 1917 his use of Irving Berlin's Tin Pan Alley song, *That Mysterious Rag*, in *Parade* registered an impressive awareness of new American styles arriving in Paris; yet his paraphrase highlights features used as early as 1904 in *Le Piccadilly* and *La Diva de l'empire*, namely the syncopated rhythms and triadic melodies of ragtime. Both *Trois Petites Pièces montées*, *La Belle Excentrique*, and Satie's final two compositions, the ballets *Mercure* and *Relâche*, are based once again on march and cakewalk; in the two ballets Satie also introduces the French genre of *valse lente*, which he had mastered in the early 1900s when composing for Paulette Darty.

Mercure and *Relâche* fuelled attacks that reflected a splintering of the ranks among Satie and his young disciples. Relations became strained four months before the première of *Mercure*, when the critic Louis Laloy, an enemy of Satie's who had previously shown hostility towards Poulenc and Auric as well, wrote a glowing review of the Russian Ballet's performances of *Les Fâcheux* and *Les Biches* in Monte Carlo. Hobnobbing with the enemy was tantamount to betrayal in Satie's eyes. The 15 February 1924 issue of *Paris-Journal* contained his own sarcastic assessment of his colleagues' latest works: 'Monte-Carlo? . . . Superbly sugary . . . Various mixtures— sexual, non-sexual, and emetic . . . Lots of syrupy things . . . buckets of musical lemonade . . . Yes . . . *Les Biches* . . . *Les Fâcheux*.'

When it came time for the première of *Mercure* on 15 June 1924 at the Théâtre de la Cigale in Montmartre, Georges Auric struck back. Joined by Georges Jean-Aubry, Henri Prunières, and other French critics and musicians who had admired *Parade* and Satie's piano works of the 1910s, Auric expressed their disappointment and his own by lashing out at the 'banal repetition of café tunes in *Mercure*' and the dull, sterile quality of both *Mercure* and *Relâche*.[68] A group of surrealist poets and painters were equally vehement in their opposition and deemed Satie's music unworthy of Picasso's classical sets and costumes. As for Satie, he saw no great rift between the new ballets and his earlier one; in his view, he had approached *Mercure* and *Parade* in the same way. In an interview with the Dadaist writer Pierre de Massot, Satie explained that he wrote *Mercure* as a divertissement, a purely decorative work devoid of plot. 'The sole function of the music is to provide a sonic backdrop for Picasso's plastic poses.'[69] The statement is remarkably similar to an observation Satie had made about *Parade*: 'I composed a background for certain noises which Cocteau considers indispensable in order to fix the atmosphere of his characters.'[70]

If the use of American ragtime and the approach to both ballets were similar, why the greater hostility towards *Mercure* than *Parade*? The answer lies partly in artistic fashions. In 1917, Satie's paraphrase of *That Mysterious Rag* in *Parade* was a bold step that fostered a turning to American popular idioms and everyday themes. By 1924 the aesthetic in Paris had shifted towards neoclassicism, and composers were exploring new eighteenth-century sources for French art music. This absorption and the waning interest in American popular music prompted Auric to renounce the music-hall clichés of *Mercure*.

Auric and his critic and surrealist cohorts were probably also dismayed by the ballet's musical language. Despite Satie's descriptions of *Mercure* and *Parade* as musical backdrops, *Mercure* had a repetitive, monochromatic quality that did not characterize the earlier ballet. Satie illustrated the poignant story and the memorable characters of *Parade* with whole tone and pentatonic passages for the Chinese Magician, a Tin Pan Alley tune for the Little American Girl, dance rhythms for the Acrobats, and a chorale and a fugue. The language of *Mercure*, by contrast, resembled furniture music in its monotony. Certainly it is true that Satie could not illustrate a story since, instead of a plot, the scenario of *Mercure* contained

---

[68] Auric's reviews of *Mercure* and *Relâche* appeared respectively in 'La Musique', *Les Nouvelles littéraires* (21 June 1924), p. 7 and in *Les Nouvelles littéraires* (13 Dec. 1924), p. 7. For a discussion of the critical controversy that surrounded Satie's last two ballets, see Alan Murray Gillmor, 'Erik Satie and the Concept of the Avant-Garde', Ph.D. dissertation (Musicology), Univ. of Toronto, 1972, pp. 273–6.

[69] This is Gillmor's paraphrase of Satie's statement to Pierre de Massot. See 'Satie and the Concept of the Avant-Garde', p. 266. 'Poses plastiques' (plastic poses) is the subtitle of *Mercure* and is a music-hall term referring to gymnastic stunts performed without apparatus.

[70] See Cocteau, *The Cock and the Harlequin* in *A Call to Order*, p. 54.

three independent tableaux each depicting a separate adventure of the god Mercury. Yet instead of evoking a classical mood appropriate for adventures of the Greek gods, Satie chose to spin out endless chains of fairground and ragtime themes, waltzes, polkas and other popular dances. The utter disparity between classical mythology and ragtime and music-hall tunes was irritating, rather than amusing; the music of *Le Bœuf sur le toit*, *Les Biches*, *Le Train bleu*, and *Les Fâcheux*, if not illustrative, evoked moods appropriate for their ballet settings.

The Dada painter and author Francis Picabia was a big proponent of Satie's and defended *Mercure* with the comment, 'He dares to compose music without worrying if it will please or displease the right or the left.'[71] Picabia replied to the furious reception of *Mercure* by enlisting Satie and creating their own ballet *Relâche*, subtitled 'instantaneous ballet'. Organized by Rolf de Maré and his principal dancer Jean Borlin, *Relâche* was originally based on a piece by Blaise Cendrars called 'Après-diner', with music by Satie and sets by Picabia. At some point, however, Cendrars disappeared. Concern among the other artists mounted until, in Cendrars' absence, Picabia took command and wrote the scenario.[72] In November 1924 audiences arrived at the Théâtre des Champs-Élysées expecting to see Rolf de Maré's Swedish Ballet in a performance of the new Satie–Picabia ballet. Instead they found a closed house—true to the title *Relâche* or 'No performance'—and were told that Jean Borlin was indisposed. Following what many believed to be a trick—a fulfilment of the title *Relâche*—the ballet opened a week later on 29 November and, according to the critic René Dumesnil, made audiences howl and heckle as they listened to Satie's inane re-creations and transpositions of the familiar 'Cadet Roussel' and other popular tunes. The uproar grew particularly wild during curtain calls when Satie drove on stage in a midget five-horsepower Citroën.[73]

The scenario of *Relâche* features a Man and a Woman, a male *corps de ballet*, dancers who smoke throughout, a fireman who wanders through the set, and costume changes occurring on stage. Action takes place against the backdrop of a wall of giant

[71] Quoted in Allan Ross MacDougall, 'Music and Dancing in Paris', *Arts and Decoration* 21 (1924): 35.
[72] For this information on the genesis of *Relâche* and on Cendrars's role, see Jay Bochner, *Blaise Cendrars: Discovery and Re-creation* (Toronto: University of Toronto Press, 1978), p. 67.
[73] Dumesnil's account appears in Roger Shattuck, *The Banquet Years*, rev. edn. (New York: Vintage Books, 1968), p. 174.

gramophone records. Satie responded with a score which is even more tedious than *Mercure*. In both ballets, but even more so in *Relâche*, new musical sections are thematically identical to previous ones but simply transposed; an opening strain is recalled more often than is typical of American marches and cakewalks; and new dance themes closely resemble melodies heard earlier in the ballet. In *Relâche*, Satie also tends to draw from a pool of cadential ideas—dotted or ragtime figures—which he pastes onto the ends of sections. The appendix (below) outlines the series of dances that appear in *Mercure* and *Relâche* and indicates where Satie employs wholesale repetition or transposition of material from a previous dance.

Picabia's intent was to be defiant, to challenge the academic, the 'Artistic', to mock his audience. '*Relâche* does not want to say anything. It is the pollen of our epoch. A speck of dust on our fingertips, and the drawing fades away . . .'[74] In a statement written on the manuscript of his ballet, Satie echoed Picabia's rebellion by heaping scorn on the 'timid' and other 'moralists' who reproached his use of popular themes. So well did Picabia feel this commentary captured the ballet's aesthetic that he made a portrait with a drawing of Man Ray's photograph of Satie at the top and Satie's ironic words superimposed above a passage from one of the *valses lentes* in *Relâche*.

A manifesto by the poet André Breton entitled 'After Dada' (dating from before 1924) described the funeral of Dada in Paris in May 1921. Breton and his friends Philippe Soupault and Paul Eluard were thankful for the end of the movement; 'its tyranny had made it intolerable,' Breton reported.[75] Picabia's *Relâche*, premièring in late fall of 1924, was thus Dada's last hurrah. Not surprisingly, the aesthetic of the ballet comprises an interesting summary of Picabia's own anarchic vision and the ideas of Dada's Rumanian founder Tristan Tzara. Picabia's article 'L'Art', appearing in Paris in February 1920, typifies such views. Here he dismissed Beauty as a visual convention which had nothing to do with life and should therefore have no bearing on art. 'Art is, and can only be, the expression of contemporary life,' he declared.[76] A manifesto 'Lecture on Dada', which Tzara wrote in 1922, similarly urged artists to recognize that life is far more interesting than art. Beauty

---

[74] Quoted by Rollo Myers in *Erik Satie* (New York: Dover Publications, 1968), p. 107.

[75] Robert Motherwell, ed., *The Dada Painters and Poets: An Anthology*, 2nd edn. (Cambridge, Mass.: The Belknap Press of Harvard University Press, 1989), p. 205.

[76] See Francis Picabia, 'Art', in Lucy R. Lippard, ed., *Dadas on Art* (Englewood Cliffs, NJ: Prentice Hall, 1971), p. 167.

in art does not exist, Tzara argued, and should be discarded in favour of an art that manifests life.[77]

This debunking of artistic conventions found its match in Satie's banal music for *Relâche* and in Picabia's inconsequential scenario which made no attempt to say anything. Yet by 1924, none of this was new. Dada in Paris had long since expired. And Dada's anti-academic beliefs and challenges of Beauty had been anticipated as early as 1913 in Satie's autobiographical sketches and paralleled by Apollinaire and Cocteau in their widely read pamphlets of 1917 and 1918, 'Parade' and *The Cock and the Harlequin*. For Poulenc, Auric, and other young French composers attending the première of *Relâche* in 1924, the ballet's ideas seemed trite. *Relâche* was a period piece which looked back, rather than inspiring future activity as *Parade* had done. In its conception as a Dada ballet, however, it offered a more successful integration of music, dance, and scenario than *Mercure*. Further, it contained the first use of film in a ballet. *Entr'acte* by René Clair with music by Satie was presented during the ballet's intermission and judged by such Dada writers as Georges Hugnet 'one of the finest films I know, hardly surpassed by those which Dali and Man Ray were to make later on'.[78] Clearly *Entr'acte* was the single avant-garde element.

After *Relâche*, Satie ceased composing. He retained the staunch loyalty of Darius Milhaud, Jean Wiéner, and a new group of musicians, mostly introduced by Milhaud, who dubbed themselves the 'Arcueil School' after Satie's suburb. Their interests coincided little with 'Les Six' and did not embrace the music-hall. The collaborative activity spawned by café-concert, fair, and music-hall thus came to a close in late 1924 with the productions of *Mercure* and *Relâche*. Satie's death on 1 July 1925 clinched the end of an era. Yet, as hackneyed as Satie's last two ballets may have sounded to musicians in the mid-1920s, both held elements of the avant-garde spirit that had transformed French music in the late 1910s. The example set by Satie's experimentation with vernacular idioms and by his radical blend of popular and art music proved fundamental to the styles of Auric, Milhaud, and especially Poulenc throughout their musical careers. Auric's *Les Matelots* (The Sailors) (1925) and his film scores, Milhaud's *Six chants populaires hébraiques* (1925) and *Kentuckiana* (1948), and Poulenc's *Chansons gaillardes* (Jolly Songs) (1926) and *Banalités* (1940)—each continued to discard a subjective, rarefied art in favour of a depersonalized language that drew on the musical sounds of everyday life.

[77] Motherwell, ed., *The Dada Painters and Poets*, p. 248.
[78] Ibid., p. 192.

Epilogue

AN article in the Sunday *New York Times* (1985) devotes its space entirely to a discussion of a score by William Bolcom, *Songs of Innocence and of Experience*, which calls for soloists, choruses, orchestra, and rock, folk, and jazz instrumentalists. Such a massive array of performing forces is necessary in order to bring alive the large number of styles and idioms assembled in Bolcom's three-hour song cycle. Complex chromatic passages contrast with folk ballads, country waltzes, and jazz rock. In the words of John Rockwell, writing one year after the première, the score epitomizes one of the 'most talked about trends today, eclecticism'.[1] The British scholar Arnold Whittall would substitute 'confrontation' for eclecticism but his meaning would be similar. In Whittall's view, the 'most essential contemporaneity of post-tonal music is a juxtaposition of the essences of past and present', of earlier tonal styles and an atonal language, without any attempt to link the two.[2]

Rockwell applies his term 'eclecticism' to the mixture of cultivated and vernacular styles that characterizes much of American music in the 1980s. Whittall speaks of 'confrontation' in connection with European modernists such as Alban Berg whose Violin Concerto strikingly illustrates the lack of reconciliation of tonal and chromatic languages, and with Americans such as Elliott Carter whose highly chromatic music is harmonic in orientation and uses chords (sometimes built from thirds and triads) that assume a tonic function. In the case of both eclecticism and confrontation, organic unity becomes an inappropriate gauge for analysis.[3]

Yet organic unity was an unsuitable gauge long before William Bolcom and Elliott Carter, and for similar reasons, in the music composed by Satie, Milhaud, Poulenc, and Auric during the 1910s and 1920s. Without moving outside a tonal language, the four composers juxtaposed both vernacular and cultivated styles and 'essences of past and present' (the assemblage of Rossini, Offenbach, and blues in *Les Biches*, of fugue and Tin Pan Alley in *Parade*, of

[1] See John Rockwell, 'What Will Be the Fate of this Gigantic Score?', *The New York Times*, 15 Sept. 1985, p. 17.
[2] Arnold Whittall made this statement in a keynote address entitled, 'The Theorist's Sense of History: Concepts of Contemporaneity in Composition and Analysis', delivered at the joint-meeting of the AMS-SMT-SEM-CMS at Vancouver, British Columbia, on Saturday, 9 Nov. 1985.
[3] Whittall rejected 'organic unity' as a principal gauge for the analysis of post-tonal music.

folk tunes and parodies of nineteenth-century German music in *Les Fâcheux*, of Brazilian sambas and a neoclassical fugue in *Le Train bleu*), thus prefiguring a central development in late twentieth-century music.

Certainly Satie and his circle were not alone in the 1920s in their use of popular sources. Stravinsky, Debussy, and Ravel also incorporated French and American popular material. In the case of Satie, Milhaud, Poulenc, and Auric, however, the approach to vernacular idioms involved a use of principles antithetical to the nineteenth-century German notion of a sublime, well-integrated work: principles such as diversity and simultaneity of material, repetition, satire. Curiously it is only today, when these techniques have permeated late twentieth-century music and composers are juxtaposing heterogeneous musical sources, that we can look back on the works of Satie and his circle and apply appropriate analytic criteria. Thus the present illuminates the past. Satie's *Parade*, performed by the Joffrey Ballet or at the Metropolitan Opera, has never sounded so contemporary.

Appendix

Mercure

Marche-Ouverture

Premier Tableau

La Nuit
Danse de tendresse
Signes de Zodiaque
Entrée de Mercure (Transposition and literal recall of 'Marche-Ouverture')

Deuxième Tableau

Danse des Graces
Bain des Graces (Transposition and rhythmic diminution of opening theme of 'Danse des Graces')

Troisième Tableau

Polka des lettres
Nouvelle Danse
Le Chaos (Transposition of 'Polka des Lettres')
Rapt de Proserpine

Relâche

Première Partie

Ouverturette
Projectionnette
Entrée de la femme
Musique (Transposition of parts of the 'Ouverturette')
Entrée de Borlin (Transposition and rhythmic augmentation of opening theme of 'Ouverturette')
Danse de la porte tournante
Entrée des hommes
Danse des hommes (Opening transposes 'Projectionnette')

Danse de la femme (Recall of part of the 'Danse de la porte tournante')
Final (Three statements of the folk tune, 'Cadet Roussel')

Deuxiemè Partie

Musique de rentrée (Recall of 'Entrée des hommes')
Rentrée des hommes (Transposition of 'Entrée des hommes')
Rentrée de la femme (Rhythmic Augmentation of 'Entrée de la femme')
Les Hommes se dévetissent (Recall of 'Cadet Roussel' from 'Final')
Danse de Borlin et de la femme
Les Hommes (Transposition of parts of 'Musique de rentrée')
Danse de la brouette
Danse de la couronne
Le Danseur (Transposition of parts of 'Projectionnette' and 'Ouverturette')
La Femme (Transposition of 'Entrée de la femme')
La Queue du chien (Transposition of 'Projectionnette')

Bibliography

ADRIAN, *Histoire illustrée des cirques parisiens d'hier et d'aujourd'hui*. Paris: Adrian, 1957.

ALLAIS, ALPHONSE, *Œuvres posthumes*, edited by François Caradec and Pascal Pia. 8 vols. Paris: La Table Ronde, 1966.

ANDERSON, ALEXANDRA and SALTUS, CAROL, *Jean Cocteau and the French Scene*. New York: Abbeville Press, 1984.

APOLLINAIRE, GUILLAUME, 'L'Esprit nouveau et les poètes'. *Mercure de France*, 1 Dec. 1918.

—— 'Parade', in Francis Steegmuller, *Apollinaire: Poet Among the Painters*, pp. 512–14. New York: Farrar & Strauss, 1963.

—— 'Une Œuvre nouvelle de Satie', *Littérature* 2 (1919): 23–4.

AURIC, GEORGES, 'Une Œuvre nouvelle de Satie', *Littérature* 2 (1919): 23–4.

—— 'Les Ballets-Russes: à propos de *Parade*', *La Nouvelle Revue française* 16 (1921): 224–7.

—— 'La Musique: quelques maîtres contemporains', *Les Écrits nouveaux* 9 (Mar. 1922): 70–9.

—— 'La Musique', *Les Nouvelles littéraires* (21 June 1924): 7.

—— 'La Musique', *Les Nouvelles littéraires* (13 Dec. 1924): 7.

—— 'Découverte de Satie', *La Revue musicale* 214 (1952): 119–24.

—— 'Préface', in Jean Cocteau, *Le Coq et l'arlequin*. Paris: Éditions Stock, 1979.

—— *Quand j'étais là*. Paris: Bernard Grasset, 1979.

AXSOM, RICHARD, *'Parade': Cubism as Theater*. New York: Garland Publishing, 1979.

BÉLICHA, ROLAND, 'Chronologie Satiste ou photocopie d'un original', *La Revue musicale* 312 (1978).

BERLIN, EDWARD, *Ragtime: A Musical and Cultural History*. Berkeley: University of California Press, 1980.

BERTAUT, JULES, *Les Belles Nuits de Paris*. Paris: Flammarion, 1927.

BIZET, RENÉ, *L'Époque du music-hall*. Paris: Éditions du Capitole, 1927.

BLESH, RUDI and JANIS, HARRIET, *They All Played Ragtime*. New York: Oak Publications, 1966.

BOCHNER, JAY, *Blaise Cendrars: Discovery and Re-creation*. Toronto: University of Toronto Press, 1978.

BORDMAN, GERALD, *American Musical Theatre: A Chronicle*. New York: Oxford University Press, 1978.

BOST, PIERRE, *Le Cirque et le music-hall*. Paris: René Hilsum, 1931.

BROWN, FREDERICK, *An Impersonation of Angels: A Biography of Jean Cocteau*. New York: The Viking Press, 1968.

BRUNSCHWIG, CHANTAL, *100 ans de chanson française*. Paris: Aux Éditions du Seuil, 1972.

BRUYAS, FLORIAN, *Histoire de l'opérette en France 1855–1965.* Lyon: Bruyas et Vitte, 1974.

CARADEC, FRANÇOIS and WEILL, ALAIN, *Le Café-Concert.* Paris: Atelier Hachette/Massine, 1980.

CARCO, FRANCIS, 'Au Bal-musette', *La Danse*, Dec. 1920.

CARNER, MOSCO, *The Waltz.* London: Max Parrish, 1958.

CASTLE, IRENE with DUNCAN, BOB and WANDA, *Castles in the Air.* Garden City, NY: Doubleday, 1958.

CAWS, MARY ANN, ed., *Stephane Mallarmé: Selected Poetry and Prose.* New York: New Directions Publishing Corps., 1982.

CHARTERS, SAMUEL and KUNSTADT, LEONARD, *Jazz: A History of the New York Scene.* Garden City, NY: Doubleday, 1962.

Chat Noir, Le, Sept. 1887–Mar. 1895. Geneva: Slatkin Reprints, 1972.

CHEPFER, GEORGES, 'La Chansonette et la musique au café-concert', in Ladislas Rohozinski, ed., *Cinquante ans de musique française de 1874 à 1925.* 2 vols. Paris: Librairie France, 1925, ii. 222–70.

CHILTON, JOHN, *Who's Who in Jazz: Storyville to Swing Street.* New York: Time-Life Records Special Edn., 1978.

COCTEAU, JEAN, '*Parade*: Ballet réaliste: In Which Four Modernist Artists Had a Hand', *Vanity Fair* 9 (Sept. 1917).

—— 'La Jeunesse et la Scandale', in *Œuvres complètes.* 11 vols. Geneva: Marguerat, 1946. ix. 309–45.

—— *Œuvres complètes.* 11 vols. Geneva: Marguerat, 1946.

—— *Le Rappel à l'ordre.* Paris: Éditions Stock, 1948.

—— 'Preface to *Les Mariés de la tour Eiffel*', translated by Dudley Fitts in Jean Cocteau, *The Infernal Machine and Other Plays*, pp. 153–60. New York: New Directions, 1963.

—— 'The Cock and the Harlequin', translated by Rollo Myers in *A Call to Order*, pp. 3–77. New York: Haskell House, 1974.

—— *Portraits-souvenir: 1900–1914*, edited by Pierre Georgel. Paris: Librairie générale française, 1977.

CŒUROY, ANDRE, 'Further Aspects of Contemporary Music', *The Musical Quarterly* 15 (1920): 547–73.

—— *La Musique française moderne: quinze musiciens français.* Paris: Librairie Delagrave, 1922.

—— *Panorame de la musique contemporaine.* Paris: Les Documentaires, 1928.

COLLAER, PAUL, *A History of Modern Music*, translated by Sally Abeles. Cleveland: The World Publishing Co., 1961.

—— 'I "Sei": Studio del'evoluzione della musica francese dal 1917 al 1924'. *L'Approdo musicale* 19–20 (1965): 11–78.

—— *Darius Milhaud.* Geneva: Éditions Slatkine, 1982.

COLLET, HENRI, 'Un livre de Rimsky et un livre de Cocteau: les cinq russes, les six français, et Erik Satie'. *Comoedia* (16 Jan. 1920): 2.

—— 'Les "Six" français'. *Comoedia* (23 Jan. 1920): 2.

CONTAMINE DE LATOUR, J. P., 'Erik Satie intime: souvenirs de jeunesse', *Comoedia* (3, 4, 6 Aug. 1925): 4.

Cook, Will Marion, 'Clorindy, The Origin of the Cakewalk', in *Anthology of the Afro-American in the Theater*, edited by Lindsay Patterson, pp. 51–5. Cornwell Heights, Pennsylvania: The Association for the Study of Afro-American Life & History, 1976.

Cooper, Martin, *French Music from the Death of Berlioz to the Death of Fauré*. London: Oxford University Press, 1951.

Coq, Le. [later *Coq parisien, Le*], Nos. 1–4 (May–Nov., 1920).

Courrier Musical, Le, Mar. 1918.

Craft, Robert, ed., *Stravinsky: Selected Correspondence*. Vol. i. New York: Alfred A. Knopf, 1982.

Cuneo-Laurent, Linda, 'The Performer as Catalyst: The Role of the Singer Jane Bathori in the Careers of Debussy, Ravel, "Les Six" and Their Contemporaries in Paris, 1904–1926'. Ph.D. dissertation (Music), New York University, 1982.

Daniel, Keith, *Francis Poulenc: His Artistic Development and Musical Style*. Ann Arbor: UMI Research Press, 1982.

Darlington, Marwood, *Irish Orpheus. The Life of Patrick S. Gilmore, Bandmaster Extraordinary*. Philadelphia: Olivier-Manoy-Klein, 1950.

Davies, Laurence, *The Gallic Muse*. London: J. M. Dent & Sons, 1967.

Dreyfus, Robert, *Petite Histoire de la revue de fin d'année*. Paris: Fasquelle, 1909.

Dumesnil, René, *La Musique contemporaine en France*. 2 vols. Paris: Librairie Armand Colen, 1930.

—— *La Musique en France entre les deux guerres 1919–1939*. Geneva: Éditions du milieu du monde, 1946.

Erismann, Guy, *Histoire de la chanson*. Paris: Pierre Waleffe, 1967.

Feschotte, Jacques, *Histoire du music-hall*. Paris: Presses universitaires de France, 1965.

Figaro, Le, July 1878, 1900–20, 1922.

Fréjaville, Georges, *Au music-hall*. Paris: Aux éditions du monde nouveau, 1923.

Fuld, James J., *The Book of World-Famous Music: Classical, Popular, & Folk*. Rev. edn. New York: Crown Publishers, 1971.

Garnier, Jacques, *Forains d'hier et d'aujourd'hui*. Orléans: Les Presses, 1968.

Gaulois, Le, 1900–20.

Gil Blas illustré, 1891–5.

Gillmor, Alan Murray, 'Erik Satie and the Concept of the Avant-Garde'. Ph.D. dissertation (Musicology), University of Toronto, 1972.

—— *Erik Satie*. Boston: Twayne Publishers, 1988.

Goddard, Chris, *Jazz Away from Home*. New York: Paddington Press, 1979.

Goffin, Robert, *Aux frontières du jazz*. 3rd edn. Paris: Les Documentaires, 1932.

—— *Jazz: From the Congo to the Metropolitan*, trans. Walter Schaap and Leonard G. Feather. New York: Da Capo Press, 1975.

Goléa, Antoine, *Georges Auric*. Paris: Ventadour, 1958.

GOWERS, WILLIAM PATRICK, 'Erik Satie: His Studies, Notebooks and Critics'. Ph.D. dissertation (Musicology), Cambridge University, 1965.

GUSHEE, LAURENCE, Liner Notes to *Steppin' on the Gas: Rags to Jazz 1913–1927*. New World Records, 269.

HAMM, CHARLES, *Yesterdays: Popular Song in America*. New York: W. W. Norton, 1979.

—— *Music in the New World*. New York: W. W. Norton, 1983.

HARDING, JAMES, *Erik Satie*. New York: Praeger, 1975.

HAUGHTON, JOHN ALAN, 'Darius Milhaud: A Missionary of the Six.' *Musical America* 27 (13 Jan. 1923): 3, 42.

HENRY, LEIGH, 'We are Seven'. *Modern Music* 1, No. 2 (June 1924): 10–17.

HERBERT, MICHEL, *La Chanson à Montmartre*. Paris: Éditions de la Table Ronde, 1967.

HEYMAN, BARBARA, 'Stravinsky and Ragtime'. *The Musical Quarterly* 68, No. 4 (1982): 543–62.

HIRSBRUNNER, THEO, *Debussy und seine Zeit*. Berlin: Laaber-Verlag, 1981.

HODEIR, ANDRÉ, *Jazz: Its Evolution and Essence*, translated by David Noakes. New York: Grove Press, Inc., 1956.

International Musicological Society, 'Critical Years in European Musical History, 1915–1925', *Report of the Tenth Congress* (1967): 216–47.

JACQUES-CHARLES, 'Naissance du music-hall', in *Les Œuvres libres*, 204 (Nov. 1952): 97–164.

—— *Cent ans de music-hall: histoire générale du music-hall de ses origines à nous jours*. Geneva: Éditions Jeheber, 1956.

JANIS, ELSIE, *So Far, So Good!* New York: E. P. Dutton & Co., 1932.

KIMBALL, ROBERT and BOLCOM, WILLIAM, *Reminiscing with Sissle and Blake*. New York: The Viking Press, 1972.

KNAPP, BETTINA and CHIPMAN, MYRA, *That was Yvette: The Biography of Yvette Guilbert, the Great Diseuse*. New York: Holt, Rinehart & Winston, 1964.

LALOY, LOUIS, 'A la Comédie des Champs-Élysées'. *Comoedia* (23 Feb. 1920): 1–2.

LANDORMY, PAUL, *La Musique française après Debussy*. Paris: Gallimard, 1943.

LIPPARD, LUCY R., ed., *Dadas on Art*. Englewood Cliffs, NJ: Prentice Hall, 1971.

MacDOUGALL, ALLAN ROSS, 'Music and Dancing in Paris', *Arts and Decoration*, 21 (1924).

MALLARMÉ, STÉPHANE, 'The Book: A Spiritual Instrument', trans. Mary Ann Caws, in *Stéphane Mallarmé: Selected Poetry and Prose*, edited by Mary Ann Caws, pp. 80–4. New York: New Directions Publishing Corps., 1982.

MARINETTI, F. T., 'The Variety Theatre', translated by Umbro Apollonio, in *The Documents of Twentieth Century Art*, edited by Umbro Apollonio, pp. 126–31. New York: The Viking Press, 1970.

MARNOLD, JEAN, 'Un Spectacle d'avant-garde: avant *Le Bœuf sur le toit* de M. Jean Cocteau'. *Comoedia* (21 Feb. 1920): 1–2.

—— 'Revue de la quinzaine'. *Mercure de France* (15 Apr. 1920): 782–91.
MAYR, W., 'Entretien avec Erik Satie'. *Le Journal littéraire*, 14 (1977): 11.
MILHAUD, DARIUS, 'The Evolution of Modern Music in Paris and in Vienna'. *North American Review* 217 (Apr. 1923): 544–54.
—— 'Les Ressources nouvelles de la musique'. *Esprit nouveau* 25 (1924): n.p.
—— *Études*. Paris: Éditions Claude Aveline, 1927.
—— *Entretiens avec Claude Rostand*. Paris: René Julliard, 1952.
—— *Notes without Music*, translated by Donald Evans, edited by Rollo Myers. London: Dennis Dobson, 1952.
MOTHERWELL, ROBERT, ed., *The Dada Painters and Poets: An Anthology*. 2nd edn. Cambridge, Mass.: The Belknap Press of Harvard University Press, 1989.
MYERS, ROLLO, *Erik Satie*. New York: Dover Publications, 1968.
OSTRANSKY, LEROY, *Jazz City*. Englewood Cliffs, New Jersey: Prentice-Hall, 1978.
POULENC, FRANCIS, *Entretiens avec Claude Rostand*. Paris: René Julliard, 1954.
—— *Journal de mes mélodies*. Paris: La Société des amis de Francis Poulenc chez Grasset, 1964.
—— *Correspondance 1915–1963*, compiled by Hélène de Wendel. Paris: Éditions du Seuil, 1967.
—— and AUDEL, STÉPHANE, *My Friends and Myself*, translated by James Harding. London: Dennis Dobson, 1978.
POUND, EZRA, 'On the Swings and Roundabouts: The Intellectual Somersaults of the Parisian vs. the Londoner's Effort to Keep his Stuffed Figures Standing', *Vanity Fair* 18 (Aug. 1922): 49.
PRADER, JACQUES, 'Les Maries . . . du Groupe des Six'. Liner notes to *Les Mariés de la tour Eiffel*. Ades 14.007.
PRATT, GEORGE C., *Spellbound in Darkness: A History of the Silent Film*. Rev. edn. Greenwich, Conn.: New York Graphic Society, Ltd. 1973.
PRUNIÈRES, HENRI, 'Francis Poulenc'. *The Sackbut* 8 (1928): 189–93.
'Quand la France découvre le jazz', Special issue of *Jazz*, 325 (Jan. 1984).
REARICK, CHARLES, *Pleasures of the Belle Époque: Entertainment and Festivity in Turn-of-the-Century France*. New Haven and London: Yale University Press, 1985.
RIVIÈRE, JACQUES, 'Le Sacre du Printemps', *La Nouvelle revue française* 10 (Nov. 1913): 706–30.
ROBERT, FREDERIC, *Louis Durey: l'aîné des Six*. Paris: Les Éditeurs français réunis, 1968.
ROBINSON, DAVID, *The History of World Cinema*. New York: Stein & Day, 1981.
ROCKWELL, JOHN, 'What Will be the Fate of This Gigantic Score?' *The New York Times* (15 Sept. 1985): 17.
ROLAND-MANUEL, 'Satie tel que je l'ai vu'. *La Revue musicale* 214 (1952): 9–11.
ROGERS, M. ROBERT, 'Jazz Influences on French Music'. *The Musical Quarterly* 21 (1935): 53–68.

ROSENBLUM, ROBERT, *Cubism and Twentieth-Century Art*. New York: Harry N. Abrams, 1976.

ROSENFELD, PAUL, *Musical Chronicle (1917–1923)*. New York: Harcourt, Brace & Co., 1923.

ROSENTHAL, MANUEL, *Satie, Ravel, Poulenc*. Madras, New York: Hanuman Books, 1987.

RUST, BRIAN, *Jazz Records 1897–1942*. 4th edition, revised. 2 vols. New Rochelle, NY: Arlington House Publishers, 1978.

SANOUILLET, MICHEL, *Francis Picabia et '391'*. 2 vols. Paris: Eric Losfeld, 1966.

SATIE, ERIK, *The Writings of Erik Satie*. Translated and edited by Nigel Wilkins. London: Eulenburg, 1980.

—— *Écrits*, edited by Ornella Volta. Paris: Éditions Champ Libre, 1981.

—— *Satie Seen Through His Letters*, edited by Ornella Volta, trans. by Michael Bullock, intro. by John Cage. London, New York: Marion Boyais, 1989.

SCHULLER, GUNTHER, *Early Jazz: Its Roots and Musical Development*. New York: Oxford University Press, 1968.

—— *Musings: The Musical Worlds of Gunther Schuller*. New York: Oxford University Press, 1986.

SHATTUCK, ROGER, *The Banquet Years: The Origins of the Avant-Garde in France, 1885 to World War I*. Rev. edn. New York: Vintage Books, 1968.

SOUSA, JOHN PHILIP, *Marching Along: Recollections of Men, Women and Music*. Boston: Hale, Cushman & Flint, 1928.

SOUTHERN, EILEEN, *The Music of Black Americans*. 2nd edn. New York: W. W. Norton, 1983.

—— ed., *Readings in Black American Music*. 2nd edn. New York: W. W. Norton, 1983.

SPAETH, SIGMUND, *A History of Popular Music in America*. New York: Random House, 1948.

STEARNS, MARSHALL, *Jazz Dance: The Story of American Dance*. New York: MacMillan, 1968.

STEEGMULLER, FRANCIS, *Cocteau: A Biography*. Boston: Little, Brown & Co., 1970.

STEIN, CHARLES, ed., *American Vaudeville as Seen by its Contemporaries*. New York: Alfred Knopf, 1984.

STRAVINSKY, IGOR, *An Autobiography*. New York: W. W. Norton, 1936.

STUCKENSCHMIDT, H. H., *Twentieth-Century Music*, translated by Richard Deveson. New York: McGraw-Hill, 1969.

TAPPOLET, WILLY, *Arthur Honegger*. Zurich: Atlantic Verlag, 1954.

TEMPLIER, PIERRE-DANIEL, *Erik Satie*. Paris: Rieder, 1932.

THOMSON, VIRGIL, *A Virgil Thomson Reader*. Boston, Mass.: Houghton Mifflin Co., 1981.

Vanity Fair, 1917, 1922.

Variety Magazine, 1917–21.

VERNILLAT, FRANCE and CHARPENTREAU, JACQUES, *Dictionnaire de la chanson française*. Paris: Larousse, 1968.

VOLTA, ORNELLA, ed. *L'Album des Six*. Exhibition Catalog, Paris: Éditions du Placard, 1990.

VUILLERMOZ, ÉMILE, *Musique d'aujourd'hui,* 4th edn. Paris: Les Éditions G. Crès and Co., 1923.

—— 'The Legend of the Six', *Modern Music* 1, No. 1 (Feb. 1924): 15–19.

WEHMEYER, GRETE, *Erik Satie*. Regensburg: Gustav Bosse Verlag, 1974.

WHITING, STEVEN MOORE, 'Erik Satie and Parisian Musical Entertainment, 1888–1909', Master's thesis (Musicology), University of Illinois, Urbana, 1983.

WIÉNER, JEAN, *Allegro appassionato*. Paris: Pierre Belfond, 1978.

WILSON, EDMUND, 'The Aesthetic Upheaval in France: The Influence of Jazz in Paris and Americanization of French Literature and Art', *Vanity Fair* 17 (Feb. 1922): 49, 100.

WITMARK, ISIDORE and GOLDBERG, ISAAC, *The Story of the House of Witmark: From Ragtime to Swingtime*. New York: Lee Furman, Inc. 1939.

WRIGHT, GORDON, *France in Modern Times: From the Enlightenment to the Present*. New York: W. W. Norton, 1981.

ZELDIN, THEODORE, *France 1848–1945: Taste and Corruption*. Oxford: Oxford University Press, 1980.

Index

Page numbers in italic indicate references to sections.

Adieu, New York! (Auric) 172, 173–5, 176
Adrian 27, 28, 29, 39, 90
Alexander's Ragtime Band (Irving
  Berlin) 55, 61
Allais, Alphonse 66, 81–3
Alphabet (Auric) 163
Apollinaire, Guillaume 1, 2, 8–9, 114–15,
  143, 172–3
Arcueil School 210
Arnold, Billy 91
At a Georgia Campmeeting (Kerry
  Mills) 75–6, 77, 78
Audel, Stéphane 3, 14, 16, 86, 99, 100, 191
Auric, Georges 1, 2, 6, 14, 15, 16, 17, 19, 86,
  91, 92, 97, 100–2, 105, 106, 107, 108,
  109, 153, 171, 190, 206–7, 210, 211–12
  Adieu, New York! 172, 173–5, 176
  Alphabet 163
  Les Fâcheux 106, *198–201*
  Huit Poèmes de Jean Cocteau 161–6
  "*Ouverture*" (*Les Mariés de la tour Eiffel*)
  188
Axsom, Richard 15, 89, 112, 113, 114, 172

"Baigneuse de Trouville, La" (Poulenc, from
  Les Mariés de la tour Eiffel) 190
Bathori, Jane 3
Belicha, Roland 67, 72
Bechet, Sydney 59
Berger, Rodolphe 73
Berlin, Edward 48, 49, 75, 76, 174, 205
Berlin, Irving:
  Alexander's Ragtime Band 55, 61
  Stop! Look! Listen! 52
  That Mysterious Rag 122, 132–3, 136–42
Bertaut, Jules 25
Bertin, Pierre 65, 175
Bestiaire, Le (Poulenc), *see* "Le Dauphin"
Biches, Les (Poulenc) 146, *191–8*
Black Swan Company 96
Blake, Eubie:
  Shuffle Along 96
Blesh, Rudi 47, 48, 49
Bloch, Jeanne 87
Bochner, Jay 1, 2, 208
Boeuf sur le toit, Le (Milhaud) 31, 32, 44,
  172, 173, 182, 183–6
Bolcom, William 56, 210

Bordman, Gerald 51, 52
Bost, Pierre 29
Braque, Georges:
  Still Life with Violin and Pitcher 8
Breton, André 209
Brooks, Shelton:
  The Darktown Strutters' Ball 54, 62
Brown, Frederick 4, 19
Brunschwig, Chantal 73
Bunch o' Blackberries (Abe Holzmann) 49
Burleigh, Harry T. 95

Caradec, François 24, 25, 26, 33, 34, 36, 40,
  44, 87, 88
Caramel mou (Milhaud) 174–7
Carco, Francis 87
Carner, Mosco 73
Casino de Paris 19, 34, 40, 52, 59
Castle, Vernon and Irene 54–6, 61, 63
Cendrars, Blaise 1, 2, 5, 43, 97, 202, 208
Chaplin, Charlie 43–4, 89, 183
Charpentreau, Jacques 25, 87
Charters, Samuel B. 57, 59, 95
Chat Noir, The 20–3, 66–7, 81
"Chemise, La" (Satie) 80
Chepfer, Georges 25
Chevalier, Maurice 40, 87–8, 180
Chilton, John 58, 59
Chocolat 28, 39, 49
Christiné, Henri 87, 196
  Entr'acte (René Clair) 210
Cinéma-Fantaisie 101
Cinématographe 41, 42
Clair, René 93, 210
Claudel, Paul 89, 90
Clorindy; or, The Origin of the Cakewalk
  (Harrigan and Hart) 48
Cocardes (Poulenc) 100, 146, 172, 173, 177–
  81
Cock and the Harlequin, The, 4, 7–11, 14,
  88, 106–9
Cocteau, Jean 1, 2, 4–6, 12, 13, 18, 27, 30,
  35–6, 37, 39, 43, 86, 88–9, 91, 92, 93, 94,
  98, 110, 171–3, 178, 197
  Le Boeuf sur le toit 31
  The Cock and the Harlequin 7–11, 14,
  106–9
  Les Mariés de la tour Eiffel 186–7

Cocteau, Jean (*cont.*)
  Parade 112–15
  "Toréador" 153–4
  Le Train bleu, 198
Cohan, George M.:
  Over There 54
Collaer, Paul 100, 101, 163, 177, 198, 201
Collet, Henri 5
Columbia Records 63
Confrey, Zez:
  Kitten on the Keys 176
Contamine de Latour, P. 66
Cook, Will Marion, 59, 91
  Clorindy; or, The Origin of the Cakewalk
  48
  In Dahomey 48
Création du monde, La (Milhaud) 97, *201–5*
Cubism 8
Cuneo-Laurent, Linda 3

Dada 66–7, 81, 149, 151, 178–9, 181, 208–10
Damia 105
Daniel, Keith 2, 5, 87, 91, 98, 154, 160, 190,
  196
Darktown Strutters' Ball, The (Shelton
  Brooks) 54, 62
Darlington, Marwood 46
Darty, Paulette 37, 64, 72–3, 74, 79–80
"Dauphin, Le" (Poulenc, from *Le Bestiaire*)
  168–70
Debussy, Claude 10, 11, 12, 17, 28, 75, 130,
  144
Debussyites, Debussyism 13, 14–16, 100,
  101, 175
Deiro, Pietro 63
Dépaquit, Jules 80
Deslys, Gaby 40, 52–3, 90–1
Descriptions automatiques (Satie) 166
Devidons la bobine (Dominique Bonnaud,
  Numa Blès) 74
Diaghilev, Serge 13, 191
"Dîner à l'Élysée, Un" (Satie) 68–72
"Discours du Général" (Poulenc, from *Les
  Mariés de la tour Eiffel*) 188–90
"Diva de l'Empire, La" (Satie) 72, 74–8, 79
Donnay, Maurice 21, 66
Dranem 26
Dreyfus, Robert 35
Dumesnil, René 17, 208
Dunbar, Paul Laurence:
  Clorindy; or, The Origin of the Cakewalk
  48
  In Dahomey 48
Durey, Louis 1, 2, 4, 6

Edison Kinetoscope 41
Embryons desséchées (Satie) 181

Enfantillages pittoresques (Satie, from
  Enfantines) 22, 163, 169
Erismann 21, 23, 25, 26, 33
Europe, James Reese 54, 56–8
  The Castle Doggy 62
  Congratulations Waltz 62
  Monkey Doodle 62
Europe's Society Orchestra 56, 57, 63
Everybody's Doing it Now 55
Everybody Shimmies Now 176

Facheux, Les (Auric) 106, *198–201*
de Feraudy, Maurice 73, 79
Feschotte, Jacques 32, 33, 34, 35, 36, 40, 41,
  42, 44, 172
Fletcher, Jimmy 53
Folies-Bergère, Les 36, 42, 44
Foottit 28, 39, 49, 172
Fratillini clowns 28, 90, 104, 172
Fréjaville, Georges 28, 31, 32, 38, 41, 42
Fuller, Loie 35
Fursy, Henri 23–4

Garnier, Jacques 29, 113
Gaya, Bar 92
"Géneral Lavine–eccentric" (Debussy, from
  Préludes, Book II) 28
Gillmor, Alan Murray 207
Gilmore, Patrick S. 46–7
Goffin, Robert 53, 58, 59, 93, 176
Goldberg, Isaac 60, 61
Goudeau, Émile 21
Gowers, William Patrick 79
Gramophone Company 63
Grock 28, 90
Guiter, Jean-Paul 55, 63

Hamm, Charles 47, 48, 51, 55, 56, 57, 58
Handy, W. C.:
  Memphis Blues 56, 165
  St Louis Blues 93, 204
Harding, James 74, 192
Harrigan, Edward "Ned":
  Walking for dat Cake 48
  Cordelia's Aspirations 48
Hart, Tony:
  Walking for dat Cake 48
  Cordelia's Aspirations 48
Haughton, John Alan 15, 16, 104
Held, Anna 51
Hellfighters, The 57–8
Herbert, Michel 20, 21, 22, 23, 24, 38
Henry, Leigh 5
Heyman, Barbara 165
Hodeir, André 203, 204
Hogan, Ernest:
  Clorindy; or, The Origin of the Cakewalk
  48

Holzmann, Abe:
    Bunch o' Blackberries 49
    Hunky Dory 77
Honegger, Arthur 1, 2, 3, 4, 6
Hugnet, Georges 210
Huit Poèmes de Jean Cocteau (Auric) 161–6
Hunky Dory (Abe Holzmann) 77
Huyghens, Salle 2, 3
Hyspa, Vincent 22, 79–80
    Chansons d'humour 68
    "Un Dîner à l'Élysée" 68–72
    "Tendrement" 72, 74
Impressionism, and anti-Impressionism 9–17
In Dahomey 48
Indiens Sioux, Les 27
Italian Futurism 7–8

Jack in the Box (Satie) 31
Jacob, Max 1, 2, 5
Jacques-Charles 33, 35, 40, 44, 52, 54
Janis, Elsie 53–4
Janis, Harriet 47, 48, 49
Jazz Kings, The 53, 59, 64, 88, 90, 176
"Je te veux" (Satie) 72
Johnson, James P.:
    Old-Fashioned Love 93
Jouy, Jules 66
Joyeux Nègres 49–50

Kildare, Walter 59
Kimball, Robert 56, 96
Kitten on the Keys (Zez Confrey) 176
Kochno, Boris 198
Krenek, Ernst 185
Kunstadt, Leonard 57, 59, 95

Laisse-les-tomber 40, 52–3, 59, 88
Laloy, Louis 172, 206
Landormy, Paul 99, 100
Léger, Fernand 202, 203
Lemaître, Jules 24
Little Tich 35
Liza (Maceo Pinkard) 96
Lowry, Vance 93, 204
Lumière, Louis and Auguste 41

MacDougall, Allan Ross 208
Mallarmé, Stéphane 8, 9
"Marche nuptiale" (Milhaud, from *Les Mariés de la tour Eiffel*) 188
de Maré, Rolf 153, 202, 208
Mariés de la tour Eiffel, Les (Milhaud, Poulenc, Auric) 5, 30, 35–6, 153, *186–90*
Marinetti, F. T. 8
Marnold, Jean 173
Mayol 99, 161

Mayr, W. 1
Médrano, Cirque 27, 28, 90
Memphis Blues (W. C. Handy) 56, 165
Menus propos enfantins (Satie, from *Enfantines*) 28, 163, 170
Mercure (Satie) 206–8
Messager, André 195, 196
Meyer, Marcelle 93
Meyers, Joe 59
Milhaud, Darius 2, 3, 4, 6, 9, 14–16, 19, 43, 86, 89–92, 94–7, 101, 102–6, 107, 109, 110, 153, 170, 190, 191, 197, 210, 211–12
    Le Boeuf sur le toit 31, 32, 44, 172, 173, 181, 183–6, 206
    Caramel mou 174–7
    La Création du monde 201–5
    "Marche nuptiale" (*Les Mariés de la tour Eiffel*) 188
    Le Train bleu 198–201
    Trois Poèmes de Jean Cocteau 166–8
Milhaud, Madeleine 95
Mills, Kerry 48
    At a Georgia Campmeeting 75–6, 77, 78
Mistinguett 40, 88, 109, 161
Mitchell, Louis 58–9
Motherwell, Robert 67, 209, 210
Mouvements perpétuels (Poulenc) 3, 170–1
Myers, Rollo 209

Nevers, Daniel 63
Nichols, Roger 192
Les Noces (Stravinsky) 192
Nordica, Lillian 46
Norton, Lilian, *see* Nordica, Lillian
Nouveau Cirque, Le 27–9, 39, 49–50
"Nouveaux Jeunes, Les" 1–3, 4, 5
Novelty Jazz Band, The 91

Offenbach, Jacques 195, 196, 200, 201
Old-Fashioned Love (James P. Johnson) 93
Ossman, Sylvester 63
Ostransky, Leroy 95
"Ouverture" (Auric, *Les Mariés de la tour Eiffel*) 188
Over There (George M. Cohan) 54

Parade (Satie) 1, 2, 4, 6, 14, 16, 67, 100, 102, 105–6, *112–52*, 173
    "Steamship Ragtime" 113, 132–7, 140–3
Parish, Dan 59
Pasler, Jann 9
Pathé Marconi 63, 64
Peccadilles importunes (Satie, from *Enfantines*) 126, 163
Petrushka (Stravinsky) 7, 184
Picabia, Francis 208, 209
Picasso, Pablo:
    Still Life with Chair Caning 8

Piccadilly, Le (Satie) 78–9
Pickford, Mary 89, 114
Piège de Méduse, Le (Satie) 67
Pilcer, Harry 40, 52–3, 88, 90–1
Pinkard, Maceo:
  Liza 96
Poulenc, Francis 2, 3, 6, 14–16, 19, 86–8, 91,
  97, 98–100, 105, 106, 107, 109, 110, 210,
  211–12
  "La Baigneuse de Trouville" (*Les Mariés
    de la tour Eiffel*) 190
  Les Biches 146, *191–8*
  Cocardes 100, 172, 173, 177–81
  "Le Dauphin" (*Le Bestiaire*) 168–70
  "Discours du Général" (*Les Mariés de la
    tour Eiffel*) 188–90
  Mouvements perpétuels 170
  Sonate 171, 173
  "Toréador" 153–61
Pound, Ezra 45
Pratt, George C. 41, 43
Prunières, Henri 186, 206
Pulcinella (Stravinsky) 181

Ragtime (Stravinsky) 165
Ravel, Maurice 67, 80, 99, 130
Raymond, Dominique 61
Relâche (Satie) 42, 84–5, 206, 208–10
Ribemont-Dessaignes, Georges 67
Ricordi Editions 62
Rite of Spring, The (Stravinsky) 9–10, 11,
  13, 17, 118
Rivière, Henri 22, 66
Rivière, Jacques 9–10, 15, 17
Robert, Frédéric 91
Robinson, David 41, 42
Rockwell, John 211
Rogers, M. Robert 63
Roland-Manuel 65, 67
Rosenblum, Robert 8
Rosenfeld, Paul 7
Rosenthal, Manuel 99
Rostand, Claude 6, 154, 160, 178, 191
Rust, Brian 62, 63

SACEM, *see* Society of Authors,
  Composers, and Publishers of Music
Sacre du printemps, Le (Stravinsky), see *Rite
  of Spring, The*
St. Louis Blues (W. C. Handy) 93, 204
Salis, Rodolphe 20, 21, 22
Sanjek, Russell 51, 60, 62
Sanouillet, Michel 149
Satie, Erik 1, 2, 3, 5, 6, 10–11, 12–16, 19,
  65–85, 93, 100, 102, 105, 106, 109, 153,
  171, 190, 211–12
  "La Chemise" 80
  Descriptions automatiques 166

"Un Dîner à l'Élysée" 68–72
"La Diva de l'Empire" 72, 74–8, 79
Embryons desséchées 181
Enfantillages pittoresques (*Enfantines*) 22,
  163, 169
"Je te veux" 72, 74
Menus propos enfantins (*Enfantines*) 28,
  163, 170
Mercure 206–8
Parade 1, 2, 4, 6, 14, 16, 67, 68, 100, 105–
  6, *112–52*, 173
Peccadilles importunes (Enfantines) 126,
  163
Le Piccadilly 78
Le Piège de Méduse 67
Relâche 42, 84–5, 206, 208–10
Socrate 170
Sports et divertissements 67, 68, 76, 166,
  170
"Steamship Ragtime" (*Parade*) 113, 132–
  7, 140–3
"Tendrement" 72, 74, 75, 79
Trois Morceaux en forme de poire 68
Trois Petites Pièces montées 172, 173, 181,
  182
Vieux Sequins et vieilles cuirasses 166
Schmitz, E. Robert 94
Schuller, Gunther 57, 95
Scotto, Vincent 87
Scrap Iron Jazzerinos, The 176
"Séance Music-Hall" 153
Seven Spades, The 176
Shattuck, Roger 1, 20, 208
Shaw, James 59
Shubert Brothers 51–2
Shuffle Along (Eubie Blake) 96
"Si fatigué" 160
Sissle, Noble 58
  Shuffle Along 96
"Six, Les" 5–6, 7
Smith, Crickett 58
Smith, Mamie 95
Snookey Ookums 55
Society of Authors, Composers, and
  Publishers (SACEM) 60
Socrate (Satie) 170
Sousa, John Philip 47, 50, 63
Southern, Eileen 58, 95, 96
Southern Syncopated Orchestra, The 59, 91
Souvenirs de mon enfance (Stravinsky), see
  *Three Little Songs: Recollections of My
  Childhood*
"Spectacle-Concert, Le" 37, 152, *171–86*
Sports et divertissements (Satie) 67, 68, 76,
  166
"Steamship Ragtime" (Satie) 113, *132–7*,
  140–3
Stearns, Marshall 48, 55, 56, 57

Steegmuller, Francis 92, 94, 110, 172
Stein, Charles 51
Still Life with Chair Caning (Picasso) 8
Still Life with Violin and Pitcher (Braque) 8
Stop! Look! Listen! (Irving Berlin) 52
Stravinsky, Igor 6–7
  Les Noces 192
  Petrushka 7, 184
  Pulcinella 181
  Ragtime 165
  The Rite of Spring 9–10, 11, 13, 17, 118
  Three Japanese Lyrics 126, 158, 160
  Three Little Songs: Recollections of My Childhood (Souvenirs de mon enfance) 126, 169, 170
  Three Pieces for String Quartet 126
Stuckenschmidt, H. H. 184, 185
Syncopating Septette, The 58

Tailleferre, Germaine 2, 3, 4, 6
Tappolet, Willy 91
Templier, Pièrre-Daniel 66, 67, 79, 170
"Tendrement" (Satie) 72, 74, 75, 79
That Mysterious Rag (Irving Berlin) 122, 133–4, *136–42*, 206
Thayer, Doris 53
Thomson, Virgil 68
Three Japanese Lyrics (Stravinsky) 126, 158, 160
Three Little Songs: Recollections of My Childhood (Souvenirs de mon enfance) (Stravinsky) 126, 169, 170
Three Pieces for String Quartet (Stravinsky) 126
Too Much Mustard 55, 63
Train bleu, Le (Milhaud) *198–201*
"Toréador" (Poulenc) 153–61
Trois Morceaux en forme de poire (Satie) 68, 79
Trois Petites Pièces montées (Satie) 172–3, 181, 182

Trois Poèmes de Jean Cocteau (Milhaud) 166–8
Tzara, Tristan 209–10

Van Eps, Fred 63
Véritables Préludes flasques (Satie) 83
Variety Theatre, The (Marinetti) 7
Vernillat, France 25, 87
Victor Recording Company 62–3
Vieux Sequins et vieilles cuirasses (Satie) 166
Volta, Ornella 1, 3, 13, 64
Vuillermoz, Émile 5, 17

Walker, George:
  In Dahomey 48
Walters, Maurice and Florence 53
Washington, Betty 53
Waters, Ethel 95
Wehmeyer, Grete 85
Weill, Alain 24, 25, 26, 33, 34, 36, 40, 44, 87, 88
White, Eric Walter 165
Whiteman, Paul 97
Whiting, Steven Moore 22, 67, 68
Whittall, Arnold 211
Wiéner, Jean 65, 92–4, 103, 105, 210
Wilkins, Nigil 6, 67
Williams, Bert:
  In Dahomey 48
Williams, Leona 95
Wilson, Edith 95
Wilson, Edmund 45, 94
Withers, Frank 59
Witmark, Isidore 60, 61
Witmark & Sons 60–2

Yvain, Maurice 199

Zeldin, Theodore 24, 40
Ziegfeld, Florenz 51–2, 61
Ziegfeld Follies 51